T0223762

The Old and New…
A Narrative on the History of the Society for Experimental Mechanics

Synthesis SEM Lectures on Experimental Mechanics

Editor
Kristin B. Zimmerman, *SEM*

Synthesis SEM Lectures on Experimental Mechanics follow the technical divisions and the direction of The Society for Experimental Mechanics (SEM). The SEM is composed of international members of academia, government, and industry who are committed to interdisciplinary application, research and development, education and active promotion of experimental methods to: (a) increase the knowledge of physical phenomena; (b) further the understanding of the behavior of materials, structures and systems; and (c) provide the necessary physical basis and verification for analytical and computational approaches to the development of engineering solutions. The members of SEM encompass a unique group of experimentalists, development engineers, design engineers, test engineers and technicians, and research and development scientists from industry and educational institutions working in materials; modeling and analysis; strain measurement and structural testing.

The Old and New... *A Narrative on the History of the Society for Experimental Mechanics*
Cesar A. Sciammarella and Kristin B. Zimmerman
2018

Mechanics of Materials Laboratory Course
Ghatu Subhash and Shannon Ridgeway
2018

Hole-Drilling Method for Measuring Residual Stresses
Gary S. Schajer and Philip S. Whitehead
2018

The Old and New... *A Narrative on the History of the Society for Experimental Mechanics*
Cesar A. Sciammarella and Kristin B. Zimmerman

ISBN: 978-3-031-79716-3   paperback
ISBN: 978-3-031-79717-0   ebook
ISBN: 978-3-031-79718-7   hardcover

DOI 10.1007/978-3-031-79717-0

A Publication in the Springer series
SYNTHESIS SEM LECTURES ON EXPERIMENTAL MECHANICS

Lecture #3
Series Editor: Kristin B. Zimmerman, *SEM*
Series ISSN
Print 2577-6053   Electronic 2577-6088

The Old and New…
A Narrative on the History of the Society for Experimental Mechanics

Cesar A. Sciammarella

Illinois Institute of Technology, Chicago, Illinois

Edited by Kristin B. Zimmerman

SEM

SYNTHESIS SEM LECTURES ON EXPERIMENTAL MECHANICS #3

ABSTRACT

The field of Experimental Mechanics has evolved substantially over the past 100 years. In the early years, the field was primarily comprised of applied physicists, civil engineers, railroad engineers, and mechanical engineers. The field defined itself by those who invented, developed, and refined experimental tools and techniques, based on the latest technologies available, to better understand the fundamental mechanics of materials and structures used to design many aspects of our everyday life. What the early experimental mechanician measured, observed, and evaluated were things like stress, strain, fracture, and fatigue, to name a few, which remain fundamental to the field today.

This book guides you through a chronology of the formation of the Society for Experimental Mechanics, and its ensuing evolution. The Society was founded in 1935 by a very small group of individuals that understood the value of creating a common forum for people working in the field of Applied Mechanics of Solids, where extensive theoretical developments needed the input of experimental validation. A community of individuals who—through research, applications, sharp discussion of ideas—could fulfill the needs of a nation rapidly evolving in the technological field. The founders defined, influenced, and grew the field of what we now call Experimental Mechanics. Written as a narrative, the author describes, based on input from numerous individuals and personal experiences, the evolution of the New England Photoelasticity Conference to what we know today as the Society for Experimental Mechanics (SEM). The narrative is the author's perspective that invites members of the Society to contribute to the story by adding names of individuals, institutions, and technologies that have defined the Society over the past 75 years.

Many of the key individuals who greatly influenced the advancement of the field of Experimental Mechanics are mentioned. These individuals are, in many ways, the founders of the field who have written textbooks, brought their teaching leadership and experiences to the classroom, worked on the Apollo project, and invented testing, evaluation, and measurement equipment that have shaped the fields of engineering. SEM's international membership is highly represented by those in academia, as you will read, although there has always been a powerful balance and contribution from industry and research organizations across the globe.

The role of the experimental mechanician is defined, in many ways, through the individual legacies shared in the following pages....legacies that define the past and create the foundation for what is now and what is to come.

KEYWORDS

stress analysis, strain gages, photoelastic coatings, viscoelasticity, photoelasticity, Moiré, digital image correlation, experimental mechanics, materials

In Memoriam of C.E. Taylor (1924–2017),
a dear friend and colleague and, for a long time,
the Historian of SESA/SEM.
As a young student, he met and interacted with
the founding fathers of the Society,
thus becoming witness to the history of SESA/SEM for a life span
until health problems prevented him from continuing in this function.

Contents

Preface

Who are we?…

What do experimental mechanicians do for a living? *This is a good question.* How can we explain our endeavors to the public?

There are questions that through the millennia humans have posed: Why do things break?… Why do they fail? How can we prevent fractures and failures? Experimental mechanicians devote their lives to finding answers to these questions. We are the inheritors of old traditions that go back to ancient times—to the Renaissance builders of machines like Leonardo da Vinci and Galileo Galilei in the 1600s, a time that marked the transition between old philosophical thought and new scientific modern science, and tried to understand, for example, through experiment and reasoning, how beams work.

I selected a title for this book that is inspired by the founder of the Society, William M. Murray, to express in a few words the content of this account. To give a time scale of the reported events, this book covers a period beginning with the New England Photoelasticity Conference in 1935, then transforms, in 1943, into the Society of Experimental Stress Analysis (SESA), and then into its current structure the Society for Experimental Mechanics (SEM) in 1983, covering a period of 82 years. These 82 years cover an extraordinary time in the development of science and technology in the United States and worldwide. I undertook the difficult task of recognizing all the main individuals that put their efforts into creating and maintaining a great institution that has had global impact.

The true purpose of this story is to share with you the significant influence that our members have had on the definition and development of the field of Experimental Mechanics. It is my hope that you will enjoy the story and, most importantly, add to it by sharing your own stories. You will notice gaps within each legacy and there are other legacies that I'm sure we have accidentally overlooked. Therefore, if you wish to contribute additional details and stories, please send an email to director@sem.org.

Enjoy!

Cesar A. Sciammarella and Kristin B. Zimmerman
April 2018

Acknowledgments

SEM wishes to acknowledge Professor Ghatu Subhash and his student Cody Kunka, University of Florida, Gainesville, Florida, for the design and use of the SEM 75th logo on the cover.

Cesar A. Sciammarella and Kristin B. Zimmerman
April 2018

CHAPTER 1

Introduction

In this book I undertook the difficult task of recognizing all the main individuals that applied their efforts in creating and maintaining a great institution, one that has had an impact on science and technology in the U.S. and also worldwide. Since 1961, I have personally witnessed the evolution of the Society of Experimental Stress Analysis (SESA) to the current Society for Experimental Mechanics (SEM). I have attended all the meetings of the Society except for two—one in Hawaii and the other in Israel. Hence, I have met and interacted with most of the individuals that I discuss here. They are a very unique set of individuals from many different ethnicities, backgrounds, and family histories. Many of them are first-generation Americans from immigrant families or they were themselves immigrants, having escaped death and persecution from totalitarian regimes. They came to America to enjoy freedom and to live in an environment where they could fulfill their personal dreams and achievements. The coverage of their lives may vary, although I have done my best to present as balanced a protrayal as possible. The information presented in this narrative comes from the interested parties, directly or indirectly, through intricate research, and from some observations that come from my memory of facts and events.

CHAPTER 2

The Beginning...

2.1 THE NEW ENGLAND PHOTOELASTICITY CONFERENCE

Quoting W.M. Murray: "In the Spring of 1935, three (at that time) young men, J.P. Den Hartog of Harvard University, M. Lawrence Price of the Worcester Polytechnic institute, and Robert W. Vose of MIT met in Den Hartog's office for the purpose of forming a group which was called the New England Photoelasticity Conference, and which held its first meeting at MIT on June 8, 1935."

"The purpose of this group was to provide a forum for the discussion of such subjects as selection of photoelastic materials, preparation of photoelastic models, and such other topics as might be useful in connection with the application of photoelasticity to the solution of engineering problems."

"The New England Photoelasticity Conference was an immediate success. Drawing attendance from greater and greater distances, including Max M. Frocht. At the time, Max M. Frocht was an Associate Professor of Carnegie Institute of Technology, Department of Mechanics, and a former student of Professor Timoshenko at the University of Michigan." A.J. Durelli joined the conference around the same time when he was a visiting Guggenheim Fellow at MIT and joined Bill Murray's group. A.J. mentioned to me that he participated in the Photoelasticity conference since the very beginning.[1]

2.2 TRANSITION OF THE NEW ENGLAND PHOTOELASTICITY CONFERENCE TO THE SOCIETY OF EXPERIMENTAL STRESS ANALYSIS (SESA)

This phase of the evolution of our Conference covers a time span of 10 years. According to the excerpts from William M. Murray's recounts, 1933 was the year that started the process of creating a Conference to connect a group of researchers utilizing Photoelasticity as a research and practical tool for solving engineering problems. It was formalized in 1935 with the creation of the New England Photoelasticity Conference.

Let us continue by quoting again William M. Murray: "Those of us who have followed the activities of the original New England Photoelasticity Conference derive a great deal of

---

[1]Author's personal recollections.

satisfaction in seeing this informal organization continue to develop in scope and usefulness until it has grown far beyond the most optimistic hopes of its founders." He adds: "Interest spread into the general field of stress-analysis, with the result that a trend toward a wider field gradually developed. Thus, the fundamental benefits from the work of the Photoelasticity Conference extended over an increasingly broader and more important field. It soon became evident that the Conferences were pointing the way toward the formation of a new Society, national in scope and broad enough in its interests to cover the entire field of Experimental Stress Analysis." Then he states: "At the unanimous vote of the members it was decided to change the name of the organization to the Society for Experimental Stress Analysis, and to appoint a committee for the purpose of establishing the new organization."

The preceding quotes were contained in the *Proceedings of the Society of Stress Analysis* published, as Volume 1, Number 1, and constitute the final transactions of the New England Photoelasticity Conference, at its 17th semi-annual meeting and Experimental Stress Symposium.

Figure 2.1 is an image that provides a snapshot of the state of the art 75 years ago taken from a paper of William M. Murray; "An adjunct to the strain Rosette," published in the *Proceedings of the Society of Stress Analysis*, 1(1). This paper is connected to a very important development in the field of Experimental Mechanics: the advent of the electrical strain gage, one of its most important tools, the workhorse of the actual application to practically all different fields of industry and structural analysis. In Murray's era, time stress analysis was one of the fields of interest for satisfying the needs of the different classical branches of Civil Engineering, Mechanical Engineering, and Aeronautics. Confronted with the limitations of the classical semi-empirical solutions of Strength of Materials and the more advanced method of Theories of Elasticity, Plasticity, and Viscoelasticity, Photoelasticity, and different types of strain gages, became the analog methods for solving complex systems of differential equations and for verifying actual mechanical components and structures.

2.3   THE SOCIETY OF EXPERIMENTAL STRESS ANALYSIS

The Society of Experimental Stress Analysis (SESA) was born in 1943. Again, from narrative of William M. Murray in the 1943 Spring Meeting, the Executive Committee of the New England Photoelasticity Conference accepted an invitation from Charles Lipson, from the Chrysler Institute of Engineering, to hold the next Conference in the city of Detroit. The basic idea was to expand the Society to a wider spectrum of techniques. This meeting was called *The Seventeenth Semi-Annual New England Photoelasticity Conference and Experimental Stress Symposium*. In the opinion of William M. Murray: "the decision to expand the title of the Conference and, thereby, to give formal acknowledgment to the inclusion of subjects other than photoelasticity was a sound one and produced amazing results. Under Lipson's very able leadership and with financial support from the Chrysler Corporation, the local committee arranged for a three-day program which included eighteen papers. Of these, only three dealt directly with photoelasticity, whereas nine were concerned with various aspects of strain gages, two were related to

Figure 2.1: Mechanical computer to solve the strain-rosette equation. Courtesy of SESA/SEM.

brittle coatings, and the remainder covered miscellaneous applied topics." He added: "It was very apparent that henceforth the strain gage would be used for an ever-broadening range of applications which would be limited only by the imagination and ingenuity of the users. Over the years this prediction certainly became true." William M. Murray added commentaries into two important aspects of the Society's future life: (1) composition of the participants—90% were from the industry, while in the case of the New England Conference they were exclusively academics; (2) the number of participants, 256, more than twice those that attended the regular meetings of the New England Conference. These two key points are still present in our current organization. He continues: "The unusually large registration at the Experimental Stress Symposium and the fact that such a large percentage of the registrants came from industry, indicated the need for a more formal organization of the Conferences if continuity were to be maintained, and a record of the technical discussions and presentations were to be preserved."

Murray then points out another important scope of the Society, the preservation of the papers presented. At that time, he suggested asking the presenters to provide copies of their papers to distribute to the audience, a practice that lasted for many years. The other fundamental aspect of the Society that he pointed out was the creation of a dues-paying Society, a first step in providing financial support to administrative staff required to create a viable institution. He then describes how the name Society for Experimental Stress Analysis came to be. In a reunion at the house of Lipson, that included himself and Professor Miklos Hetenyi, they coined a name not tied to any technique; Society for Experimental Stress Analysis. R.D. Mindlin was asked to join the organizing group and therefore Murray, Hetenyi, Lipson, and Mindlin became the founding members of SESA.

The *17th New England Photoelasticity Conference and Symposium on Experimental Stress Analysis* took place at the Rockham Memorial Building, Detroit, MI on May 13–15, 1943. This date is considered by the Society as its starting date. For this reason, 2018 is considered the 75th anniversary of our current Society.

2.4 THE TRANSITION FROM EXPERIMENTAL STRESS ANALYSIS TO THE SOCIETY FOR EXPERIMENTAL MECHANICS AND THE FORMATION OF THE *JOURNAL OF EXPERIMENTAL MECHANICS*

The period of 1960 to today is a very remarkable time in the history of experimental methods in the U.S., Europe, and Asia. At the same time, it is a period that gave birth to a universal tool to numerically solve problems of stress analysis in the different disciplines that have been previously mentioned: the Finite Element Method (FEM). FEM practically ended the role of different analog methods for solving many complex theoretical mechanics problems. As a student of Civil Engineering at that time, I witnessed the birth of FEM as a universal tool for solving mechanics problems. It came as an outgrowth of structural mechanics and was initially "discovered" by civil engineers. In the early 1940s, designers of shell structures realized that their structural mechanics computations converged to theoretical solutions of shells. This was the spark that initiated a momentous revolution in the solution of partial differential equations that in the 1960s and 1970s obtained its actual impetus supported by large U.S. government expenditures for defense and aerospace activities. Why mention this in the history of our Society? Well, FEM methods had a profound impact in creating the basic foundations of the initial phase of our Society—solving very difficult mathematical problems by using analog methods. Some of our colleagues felt that these developments were the precursors of the demise of the Society and this recognition is at the root of the transition from the Society of Experimental Stress Analysis to the Society for Experimental Mechanics.

The first mention of the name Experimental Mechanics in our Society corresponds to the *First International Congress in Experimental Mechanics*, held November 1–3, 1961 at the New Yorker Hotel in New York City. The meeting was held with the co-sponsorship of the Office of Naval Research. The Congress was opened by D.C. Drucker, at that time president of the Society of Experimental Stress Analysis. President Kennedy sent a message of congratulations and expressed his desire that foreign participants, "unite with us in fashioning from our mutual efforts in research and technological innovation an environment in which the sinews of science can be put to work for pursuits of peace." By the way, this was the first meeting that I attended.[2]

1961 was the year that the main publication of SESA, *Journal of Experimental Mechanics*, was first published in January. Benjamin J. Lazan, one of the most active members of the

---

[2]SEM Publication: *Experimental Mechanics*, Proceedings of the First International Congress on Experimental Mechanics, New York, November 1–3, 1961, edited by B.E. Rossi.

Society in the creation of the journal, wrote an editorial in the first issue of the journal where he articulated the reasons for its creation. He pointed out that the new journal was the result of the Society's growth, the broader realm of experimental mechanics, and the role of experimental studies in verification and improvement of theory. Quoting Lazan: "the solution of the increasingly complex engineering problems must rely more and more on experimental mechanics studies to indicate limitations in current theories, to formulate more realistic and general, yet simple, assumptions required for developing new theories, and later to check the compatibility of these new theories with still newer engineering situations or simulations thereof."

It is necessary here to mention the Executive Secretary of the Society, Bonney Rossi, as one of the persons that was also a great contributor to the success of Experimental Mechanics. I remember that he edited all the papers published in the journal, providing the authors with corrections, condensations, and organization of manuscripts in his own handwriting.

The historical records refer to the society as SESA through the 1984 Fall meeting held at the Hyatt Regency Milwaukee, WI, November 4–7. The creation of the *Journal of Experimental Mechanics* was the spark that initiated the process of change of the Society from SESA to SEM. After a membership vote, the historical records indicate that the society changed the name officially in 1985. The Spring meeting that took place in Las Vegas was called the *Spring Meeting of the Society for Experimental Mechanics*.

The first Spring Conference under the name Society for Experimental Mechanics was held in 1990 in Albuquerque, New Mexico. At that time, Albert Kobayashi was SEM President. In 1993, SEM's *Spring Conference on Experimental Mechanics*, SEM's 50th Anniversary, was held at the Hyatt Regency Dearborn in Dearborn, MI, June 7–9. Frank Adams was President of the Society.

To put into historical context the timing of the several important changes in the Society, we must remember that on October 4, 1957, Sputnik 1 was successfully launched by Russia and entered Earth's orbit. This event was the starting point of the space age and the successful launch shocked the U.S. and the world. It started a period of unprecedented growth in science and technology as a response to the demands of the new technologies necessary to support new areas of engineering connected with the development of rockets, avionics, and the nuclear field, etc. These new endeavors implied great demands on material sciences, theoretical and applied mechanics, and also experimental mechanics. As one of the many participants of the Apollo project, I remember an outstanding technical event that I witnessed in the middle 1960s, where the full 42.1 m Apollo rocket system had been subjected to a compression test and fully instrumented with hundreds of strain gage bridges and strain gages…tools that are closely intertwined with the history of SEM.

2.5   REMARKABLE TECHNICAL ACHIEVEMENTS ASSOCIATED WITH EARLY TIMES OF SEM

Having gone through a chronological record of the Society's history, it is very interesting to look at the worldwide impact of its founders and their successors through important developments and inventions that have greatly contributed to what Experimental Mechanics is today.

2.5.1   BONDED STRAIN GAGES

Bonded strain gages constitute a main tool of Experimental Mechanics unparalleled for the wide impact in all branches of engineering playing a unique role in diverse types of measurement technologies. For the public at large, strain gages are utilized in simple but very important daily occurrences in many products including their utilization in many instruments in medical applications. In the structural area, mechanical and aerospace engineering strain gages play a paramount role in the certification of aircraft and aerospace structures, as has been previously mentioned, and in the development of new car models, trucks, and all sorts of earth movement equipment and lifting equipment. The variety of developments in electronic measuring devices provides the possibility of monitoring large amounts of gage bridges in real time while functioning across a large spectrum of frequencies.

When and how did strain gages came to be? There are two sources of the first use of a known property of electrical conducting materials. One relates to Edward E. Simmons, a graduate student of the California institute of Technology. The narrative of the event is that in October of 1936 Simmons suggested to his professor, Gottfried Datwayler, to utilize a constantan wire wrapped in cotton to measure the dynamic forces generated in an impact machine. This idea was implemented, and the results were presented at an ASTM meeting in June 1938. This work resulted in U.S. patent US2340146 A, filed on July 23, 1941; the inventor and assignee was Simmons.

The other development closely connected with SEM took place at the Massachusetts Institute of Technology. Quoting C.E. Taylor: "MIT was a hotbed for experimental stress analysis during the gestation period and in the infancy of the SESA." J. Hans Meier narrated his early experience as a graduate student and events connected with other successful applications of changes of resistance with deformation to measure strains. In the basement of a building at MIT, Professor Arthur Ruge had built a shake table to simulate dynamic load applied to structures. He wanted to measure the response of the simulated structures by diverse methods and finally succeeded with bonded strain gages and the instrumentation that was developed to measure the small changes of voltage generated by the utilized measuring bridges. Ruge realized the importance of the instrumentation that had been developed and Meier did his graduate work with emphasis on the electrical strain gage. Finally, all the mentioned work resulted in U.S. patent US2340146 filed on October 2, 1941, and granted on January 25, 1944 and the inventor and assignee is Arthur C. Ruge.

As is indicated by the dates of the filings of the two mentioned patents, they were essentially requested at the same time and this was a source of contention between the two inventors. According to Peter E. Stein, the conflict ended when the Company Baldwin Southwark was able to get an agreement involving both inventors and this accord was the start of the industrial manufacture of bonded resistance wire strain gages. The manufactured strain gages were given a name SR_4 Strain Gage—the S corresponded to Simmons, the R to Professor Ruge, the 4 implied two additional persons involved besides Simmons and Ruge, Dr. Donald S. Clark (a metallurgist that worked in the area of impact property of materials), and MIT Professor A.V. de Forest (who headed a group of students working in experimental stress analysis and at the same time cooperating with Professor Ruge in the manufacturing of strain gages and the instrumentation needed to record bonded strain gage outputs).

As indicated before and what C.E. Taylor called the MIT connection, due to the presence of the Society founder and first President William M. Murray, the early meetings of the Society were the forum where many of mentioned developments were presented and discussed. J. Hans Meier, was the eighth president of the Society (1950–1951). Peter K. Stein, an outstanding member of the Society who devoted his life to the strain gages field and Measurement Systems Engineering, joined while a graduate student at MIT at the time that all the above described developments were taking place; he was mentored by Professor William M. Murray.

Another devoted and important member of the MIT connection was Ferdi Stern. His connection with SESA/SEM predates the Society's history. Quoting Fred C. Bailey: "He was part of the group at MIT in the late 1930s and early 1940s that was working with de Forest and Ruge on developing the devices and techniques that would for many years predominate in the experimental stress analyst's arsenal. This led to a long affiliation with the Magnaflux Corporation, during which time Ferdi, for many of us, became Mr. Stresscoat, even though he represented the entire range of the company's products and even though he had lots of high-powered companies in the brittle coating field. He recollected with glee the spirited defenses of photoelasticity that certain prominent SESA members were driven to upon hearing the early papers describing brittle coating techniques." The other important member of the Society that formed part of the MIT connection was Greer Ellis, inventor of a commercial product, Stress-Coat, and Tenslac, and instrumentation for strain gages. He joined MIT in 1936 and for his Master's thesis he joined Professor de Forest, who assigned him a topic related to the feasibility of developing a brittle stress coating as a tool for stress analysis.

2.5.2   THE STRESS COAT

Where there is a need in the technical field human ingenuity develops a solution. In the preceding section we have seen the bonded electrical strain gage as the instrument providing reliable strain values in any mechanical and structural component. However, strain gages give a small zone answer (theoretically a point); structures have theoretically an innumerable set of points, and the points of interest are the points of threshold strain, strain that will be the source of a crack. The most practical solution was the use of the stress coat techniques to measure the larger zone of stress and strain. The first studies on the utilization of the stress coat were done in Germany by Dietrich and E. Lehr and were applied by the Maybach Company to determine principal stress directions in connecting rods and pistons of the engines of the Zeppelin air craft. The technology of stress coat was complex and the most complete study on the properties, practical use, and actual use of stress coat originated in what I will call the IIT-ITRI connection—the laboratory that A.J. Durelli founded and directed. A.J. Durelli, one of the founding fathers of SEM and one of the most versatile practitioners of Experimental Mechanics, headed a large group of researchers among them a group of young experimentalists of which I was a member. I was a witness to the research in the application of stress coatings and I remember some interesting events: stress coat was extensively used in the stress analysis of the Polaris missile, a cylindrical welded vessel subjected to internal pressure was tested; the brittle coating was sprayed and dried; the vessel was pressurized in steps. After each step, a technician encircled the crack patterns with a line and a number was added corresponding to the load level. Strain gages were added for self-calibration of the coating. At times before the actual launchings of the Polaris missile, work was done on a 24-hour basis to get results to the designers of the missile. Figure 2.2 shows the Polaris missile pressure vessel with the outlined crack lines and strain gages utilized for self-calibration. Intensive human labor was involved by very skilled technicians, and the critical points and the difference between the ideal model of the shell and the actual fabricated shell were obtained and the Polaris missile was a success. Stress coat and strain gages, originated in the MIT connection, proved to be effective tools of Experimental Mechanics. This work was reported in *Experimental Mechanics*, 1(2), Feb. 1961. A co-author was J.W. Dally, an outstanding member of the group, future president of SEM, and contributor to many aspects of the Society in addition to his remarkable scientific and technical contributions.

2.6 OUTSTANDING CONTRIBUTORS TO THE SOCIETY'S EARLY YEARS

We have presented an outline of the basic developments that were associated with the birth of the Society and the early stages of fundamental ideas that later were the seeds of the current impressive arsenal of experimental methods. It is interesting to remember some significant individuals of the past that provided the scientific underpinning of some of these developments.

William M. Murray 1910–1990.   We can state without reservation that William M. Murray played the most important role in the birth, and for many years was the driving force behind the fundamental steps in the formation of the structure, of the Society for Experimental Mechanics. He was co-founder of the Society, first President, first Secretary-Treasurer (a function that he continued to perform for many years), first honorary member, and Honorary President of SESA. Bill came to MIT in 1932 with his B.S. degree in Mechanical Engineering from McGill University. He received his M.S. and ScD degrees from MIT in 1933 and 1936, re-

Figure 2.2: Polaris missile pressure vessel with outline crack and strain gages utilized for self-calibration. (Courtesy of Cesar Sciammarella, A.J. Durelli Laboratory, IIT-ITRI.)

spectively. He became Full Professor and remained at MIT until his retirement in 1973, when he was appointed Professor Emeritus of Mechanical Engineering. After his retirement he became a Visiting Professor of Civil Engineering at University of Houston, Texas. He was a great educator in experimental stress analysis, offering the first course on photoelasticity, strain gages, and brittle coatings in a college/university curriculum. In addition, he offered the first summer sessions—one- and two-week courses for practicing engineers, scientists, and managers from industry—starting in 1953 and continuing for two decades, first at MIT and later at UCLA and in many other locations in the U.S. and Mexico. SESA was run out of his office for over a decade and his professional contributions and honors are legendary. SEM's annual Murray Lecture is awarded in his honor

Raymond D. Mindlin (1906–1987).   Mindlin was one of the great contributors to solutions of the Theory of Elasticity. His solutions are referred to as the "Mindlin Problems," a generalization of solutions of the Theory of Elasticity associated with the names Kelvin and Boussinesq. As a theoretician, he had interest in the field of photoelasticity and this interest led him to be a founding member of SESA. He received a large number of awards, most importantly the

American Presidential Medal for Merit in 1941. Another one of his important contributions to SESA/SEM was his influence on Daniel D. Drucker to write a thesis on Photoelasticity.

Charles Lipson.   Lipson was a co-founder of SESA and as a member of the Chrysler Institute of Engineering. He provided the support needed to transform the New England Photoelasticity Conference into the Society of Experimental Stress Analysis. He was also the third President of the Society. In 1968, he delivered the Murray lecture at the SEM Annual Conference.

Miklos Hetenyi (1906–1984).   Miklos received his civil engineering degree from the University of Technology and Economy of the Budapest University. He obtained his Ph.D. in Mechanical Engineering from the University of Michigan in 1936 under the guidance of Stephen Timoshenko. Upon graduation, he joined the Research Laboratory at the Westinghouse Electric Corporation. In 1946, he became a professor at the School of Engineering at Northwestern University where he developed a Photoelasticity Laboratory. In 1962, he joined Stanford University where he retired in 1972. He was one of the founders of SESA, providing input in the Society's name and in its initial organization. He was President of the Society and was made Honorary Member of the Society. In 1964, he presented the Murray lecture. The M. Hetenyi Award for the best paper published in a given year in Experimental Mechanics was given in his honor. Warren Rhines received the first M. Hetenyi Award in 1983. He was the editor of the *Society's Handbook on Experimental Stress Analysis.*

Daniel D. Drucker (1918–2001).   Daniel was one of the more remarkable members of SESA/SEM. A civil and mechanical engineer who graduated in 1940 from the University of Columbia in New York, he obtained his Ph.D. under the guidance of Raymond D. Mindlin, one of the founding fathers of SESA. His Ph.D. thesis subject was Photoelasticity, and in his thesis he developed the oblique incidence method, thus providing an optical solution to the complete determination of stresses in 2-D utilizing Photoelasticity. He was one of the outstanding applied mechanics individuals in the U.S. during the 20th century. His most important contribution to the field of applied mechanics was his concept of material stability included in his stability postulates and the Drucker-Prager yield criterion. A theoretician by trade, he had a great deal of interest in Experimental Mechanics and this interest was manifested by his association with SESA/SEM. Drucker, as a member of the Executive Committee, played an important role in the process of the creation of the Society's headquarters. As mentioned before, he also provided an important contribution to the creation of Experimental Mechanics. Drucker was President of the Society and the recipient of the Society's two highest honors, the Murray Lectureship and Honorary Membership. He also received the M. M. Frocht Award.

Thomas J. Dolan (1907–1996).   Thomas James Dolan's associations with SESA began in 1941, before the Society was formally organized, and continued with several decades of participation and service in the Society. Tom Dolan was an active and talented photoelastician that participated in the New England Semi-Annual Photoelastic Conferences. He was a member of the

Executive Committee of the *Seventeenth New England Semi-Annual Photoelastic Conference* in 1943 when the group decided to broaden its scope and become SESA. During World War II, he joined the U.S. army with the rank of captain. After returning to civilian life, he resumed his activities in SESA and was a co-author (with William M. Murray) of the chapter on photoelasticity in *SESA's Handbook of Experimental Stress Analysis*, edited by M. Hetenyi. Tom Dolan remained heavily involved in the affairs of SESA and as its ninth President from 1951–1952. In 1969, he presented the Murray Lecture and in 1975 Tom was made a Fellow and an Honorary Member.

Benjamin J. Lazan (1917–1966).   Dr. Lazan graduated with a B.S.ME from Rutgers University in 1938. He continued his education at Harvard University where he obtained a M.S. degree in Applied Mechanics. In 1939, he became an instructor in the Department of Engineering Mechanics at the Pennsylvania State University. It was at Penn State, in 1942, that he earned his doctorate and was appointed to Assistant Professor of Engineering. While at Penn State, he was in charge of research projects in the dynamic testing of materials. In 1942, Dr. Lazan became associated with the Sonntag Scientific Corp., a Baldwin Lima-Hamilton affiliate, as a project engineer. He quickly rose to chief engineer and by 1944 was an executive vice-president. While with Sonntag, Dr. Lazan developed several types of machines now used for static and dynamic testing. He was also responsible for a government research project on jet-propulsion engines. After joining Syracuse University in 1946 as an Associate Professor and Director of the Materials Laboratory, he soon became head of the Department of Materials Engineering. At Syracuse, he directed research projects for the Office of Naval Research, the U.S. Air Force, and private industry. In 1951, he went to the University of Minnesota as a Professor of Materials Engineering. Later, he became director of the engineering experimental station, head of the department of mechanics and materials, associate dean of the Institute of Technology and was, until 2017, head of the Department of Aeronautics Engineering Mechanics. He has authored over 60 papers and continued his writing endeavors until shortly before his death. Lazan played important roles during a formative time in the Society's history during which a paid Society Headquarters staff was established. One of his more important contributions to the Society was the launching of Experimental Mechanics. Dr. Lazan was a leader in the fields of dynamic testing, vibration, material damping, and fatigue. He was SESA President from 1959–1960 and delivered the Murray lecture in 1956. In 1973, the Lazan Award was created on the basis of length and quality of service through Committee activities, both in and out of official elective and appointed positions in SESA, at the local and national levels.

Francis G. "Frank" Tatnall (1896–1981).   Quoting C.E. Taylor, "He worked for the Baldwin-Southwark Corp., selling testing machines and associated equipment. He traveled widely and knew everyone there was to know in the testing business and what they were doing. In his charming and sometimes imaginative autobiography, *Tatnall on Testing* (American Society for Metals, 1966), he relates his discovery of numerous strain gages in use by imaginative inventors in various laboratories he visited. It is believed that Frank often helped these inventors set up their own businesses, had a financial interest in those, and then arranged to sell the devices through Baldwin-Southwark." In 1953, he was made the Society's first Honorary Member. An award was created in his name in 1967 for those that served the Society long and well and in 1968 Frank Tatnall received the first F.G. Tatnall Award.

Samuel Stanford Manson (1919–2013).   Samuel was born in Jerusalem and came to the U.S. at the age of 7 when his parents settled in Brooklyn, NY. He got his B.S. at Cooper Union in 1937 and in 1942 he received his M.S. from the University of Michigan. He obtained his Ph.D. from the Department of Theoretical and Applied Mechanics at the University of Illinois under the guidance of Professor JoDean Morrow. In 1942, Manson joined the Langley Research Center, a Virginia facility of the National Advisory Committee for Aeronautics, later NASA. A year later, he transferred to Cleveland's new Lewis Laboratory. In 1956, he became chief of Lewis' new Materials and Structures Division. In 1974, he retired from NASA and began teaching at Case Western Reserve University where he was a Professor of Mechanical and Aerospace Engineering and where he retired as Emeritus Professor in 1994. He is world renowned for his cutting-edge theoretical and experimental work in stress analysis, fatigue, and the development of the basic relation of low-cycle fatigue behavior (the Manson-Coffin law). He was an important contributor to the early SESA/SEM history and he served for six years on the SESA Executive Committee and was president of SESA in 1956. In 1964, he gave the Murray lecture, in 1973 he received the Lazan Award, and in 1976 he became Fellow of the Society.

D.J. DeMichele (1916–2000).   Dick DeMichele was an Honorary Member of SEM. He was active on numerous SEM committees, including the Executive Board, and served as chair of two national meetings. Dick was also the past director of the *International Modal Analysis Conference* and served as director for 13 annual IMACs. From 1940–1979, Dick DeMichele worked as a mechanical engineer at the General Electric Company in Schenectady, NY. He received the coveted Charles E. Coffin Award, the company's highest award to its employees, for his contribution in the field of solid mechanics (vibration, shock, stress-strain, and acoustics). He held U.S. patents in strain gage technology. Dick DeMichele passed away in 2000, shortly after being elected as an Honorary Member of SEM and receiving the C. E. Taylor Award.

Peter Stein (1928–2016).   We have to include another important member of the earlier times of SESA-SEM. Peter was born in Vienna in 1928. In 1938, his family escaped the Nazi regimen settled in Shanghai where they remained until 1947 when they moved to the U.S. Peter received his B.S., M.S., and Ph.D. at MIT. As graduate student at MIT working toward his Ph.D. under William Murray, Peter got the chance to be a witness to the process of the creation and development of the bonded strain gage. After a period working in industry, he joined the Arizona State University. He completed his full academic career at ASU becoming Full Professor in 1963 and retired in 1977, after which he became Emeritus Professor. After his retirement, he headed his own consulting firm. Concentrating his efforts in the general field of metrology, he developed a program in Measurements System Engineering at ASU that gained international interest. He played a fundamental role in one of the more important topics of SESA-SEM, the bonded strain gage, and was an outstanding member of the Western Regional Strain Gage Committee (WRSGC), a Technical Division of the National Society for Experimental Mechanics.

After reviewing the most important periods in the past and providing information about the most influencial individuals that shaped the SESA/SEM, let us review subsequent periods of evolution of the Society and what C.E. Taylor called connections—the interaction between individuals and institutions.

CHAPTER 3

The Illinois Institute of Technology Research Institute (IITRI) Connection

Paraphrasing C.E. Taylor …the terms "connections" and "derivatives" indicate human interdependences that allow us to follow the threads of a developing history. Two of the original members of the New England Photoelasticity Conference and SESA ended up joining the Illinois Institute of Technology (IIT): A.J. Durelli and Max M. Frocht.

Max M. Frocht was one of most famous photoelasticians of the 20th century. His two books *Photoelasticity, Volumes 1 and 2* are classics on the topic. Frocht was born in 1895 in Poland and came to the U.S. in 1912, where he moved to Detroit and got a job as a machinist and tool maker. He entered the University of Michigan in 1916 and in 1920 he received his B.Sc. in Mechanical Engineering. In the fall of 1922, he worked at the Carnegie Institute of Technology as an instructor. In 1925, he got his M.Sc. in Physics at the University of Pittsburg. He returned to the University of Michigan in 1931 where he obtained his Ph.D. His advisor was Stephen Timoshenko and the subject of his thesis was Photoelasticity. In the fall of 1931, he went back to Carnegie where he remained for the next 15 years. He organized a photoelasticity laboratory where he worked to get stress concentration factors in a variety of geometrical shapes, ultimately becaming Full Professor. In 1946, Dr. Frocht joined the Mechanics Department at the Illinois Institute of Technology and founded the Experimental Stress Analysis Laboratory.

As mentioned before, Max was one of the initial members of the New England Photoelasticity Conference and SESA/SEM and a very active participant in the meetings of the Society. He sat always in the first row and almost always gave his opinion on presented papers, often very caustically if the papers were from one of his colleagues from Chicago—at that time Frocht, Durelli, and M. Hetenyi, a Northwestern University Professor, and another of the founders of the Society. I remember a funny anecdote: Durelli was furious because Frocht pointed, correctly, to an error in a slide of Durelli's, indicating that the certain curve should be tangent to the coordinate's axis and it was not. Personally, he was always kind to me, but this behavior did not extend to other members of our group that were roasted by him with curve balls.

Frocht's lab was in one of the old buildings of IIT called Chaping Hall, a partially wooden structure where one had to climb up a steep staircase to reach the lab. The research work at the lab was done by his graduate students and by a very competent assistant of Dr. Frocht, a hold-over from Max's Carnegie times and a very able photographer, David Lansberg, who was called "mechanician" in the IIT's administrative jargon. As a Director of the Experimental Mechanics Laboratory of IIT, I inherited many things from Max, one of which was all the materials of his two photoelasticity books: page proofs, photographic negatives all carefully retouched by Lansberg with a special ink and all the neat drawings in China ink. This was a formidable piece of work done by two people, Max Frocht and his assistant David Lansberg, of course with the collaboration of many graduate students.

One of the most outstanding students of Frocht was Milton M. Leven. Quoting C.E. Taylor: "He was the ideal role model for photoelastician…The period from the late 1940s to the early 1970 was the golden era of photoelasticity." During this period, practitioners of experimental mechanics were confronted with the need to find stress distribution inside 3-D fields. An answer was found in the stress freezing method of Photoelasticity and the retrieval of information from stress-frozen specimens. Milton M. Leven, as mentioned before, excelled in the field of 3-D Photoelasticity. His work was done in the industrial environment (the Westinghouse Corporation), but also through talk courses at the Carnegie Institute of Technology. Leven was a very active and devoted member of SESA/SEM and played many different important roles in the organization of the Society. He was president of the Society in 1968 and in 1969 was nominated as an Honorary member of the Society. In 1969, he was the editor of a book: *Photoelasticity: The Selected Scientific Papers of M.M. Frocht*. Other outstanding students of Frocht were Paul Flynn and Ros Guernsey. Paul Flynn conducted photoelastic studies on stress wave propagation in solids using a rotating drum-camera that he and Lansberg had constructed. Guernsey developed algorithms for the shear difference method to separate the normal stresses at an interior point of a 3-D body using a stress frozen model and the sub-slice technique. Ros Guernsey was an important member during the formative period of the Society and was the 23rd President of the Society. Another student of Frocht was Abe Betser. Abe continued Flynn's work on wave propagation and received his Ph.D. with a thesis on this subject. Abe Betser became Professor at the Israel Institute of Technology, Haifa, Israel, attended many of SEM meetings, and published papers in Experimental Mechanics. One of the most outstanding students of Frocht was L.S. Srinath. In the 1960s, Srinath obtained his Ph.D. under Max Frocht with a thesis on diffuse light Photoelasticity. Although we were students of rival professors, we became good friends with a friendship that lasted until his death. Srinath was my host several times during a period where I frequently visited India and he invited me to give lectures at the Indian Institute of Technology in Bangalore where he was a professor. Later on, he became the President of a very important institution in India, the Indian Institute of Technology at Madras, which was a huge research center. He was the author of a book titled *Scattered Light Photoelasticity* and co-author of another book, *Experimental Stress Analysis*. He shared with me many interesting and sometimes funny stories and anecdotes about the daily life in the laboratory with Frocht. Srinath received recognition as Fellow of SEM.

Continuing with the IIT-IITRI connection and quoting one more time C. E. Taylor: "Any History of the SEM would not be complete without an article about A. J. Durelli. He was a versatile engineer, a prolific writer, and an extraordinary mentor for young experimentalists." Quoting an article that I wrote about A.J. as we used to call him, he adds: "One of his most precious legacies has been the training of many outstanding disciples, close collaborators and students. I fully agree on all accounts, especially the legacy. As the SEM Historian, I can add some facts that strongly support the allegation of Durelli's derivatives. In that group are four Murray Lecturers, three SEM Honorary Members and two SEM Past Presidents. No other educator in the history of SESA/SEM can match those numbers. The story continues because their students (second generation Durelli derivatives) have much momentum. Already they include an SEM Past President, several SEM Fellows and many SEM Award winners. The third and fourth generations of Durelli derivatives are underway and are gaining momentum."

Augusto J. Durelli was born in Buenos Aires, Argentina on April 30, 1910. He attended the School of Sciences and Engineering of the University of Buenos Aires where received a degree in Civil Engineering at the age of 22. His classmates recall him as an outstanding student, but one that did not conform to prevailing rules of behavior. Students attended school in formal attire, but not A.J. He was notorious for his neglect of dress codes, a characteristic that he preserved until the end of his life. Another one of his characteristics as a young man was his disregard for the amenities that wealth can provide. He had a Spartan sense of living, and again this was the way he lived. After his graduation, his father, a wealthy building contractor, sent him to Paris to study. In the 1930s, it was not an unusual thing for well-to-do Argentinians to go to Paris to escape the Victorian mores still prevailing in the society, however, his motivation was a quite different one. In Paris, he continued his engineering studies getting a doctoral degree at the famous Sorbonne School of the University of Paris. But this was not enough. He had a second life-lasting love: social issues. In the same year that he got his doctoral degree in Engineering, 1936, he got a doctoral degree in Social Sciences at the Catholic University of Paris. A profoundly religious person, he embraced the social activism of Jacques Maritain, a Roman Catholic philosopher. He was always on the side of the poor, the oppressed, and the needy. He published books, newspaper articles, and letters to the editor on social and political issues in Spanish, English, and French.

Upon graduation in France in 1936, he returned to Buenos Aires but he did not stay for long as he decided that becoming head of his father's construction firm was not what he wanted. He applied for a Guggenheim Fellowship, which he received and chose to join the William M. Murray group at MIT. At the end of the fellowship, he became a Visiting Professor at the Ecole Polytechnique in Montreal. He married in Montreal and had his first child. In 1944 he returned to Buenos Aires where he became the Head of the Laboratory of Testing Materials of the Municipality of Buenos Aires. Those were difficult years in Argentina, military revolt having overturned the civilian government. In October 1945, he joined a group of engineering students that seized the engineering school to protest Coronel Peron's dictatorship. A slight, short, and bald guy popped out of an external window of the Engineering building—this was the way that I met A.J. We were finally dislodged by the police and taken to jail to a huge internal patio where about 350 students were logged together. Our misfortune did not last long, as Peron was deposed and jailed and the detained students were liberated in groups. By a twist of fate, A.J. and I were left in jail in a large patio for two additional days in the most unusual circumstances that one can imagine. In 1946, he left Buenos Aires for good. He joined the recently formed Armour Research Foundation at the Illinois Institute of Technology where he became the Head of the Stress Analysis Section. In 1956, he received an additional appointment at the Illinois Institute of Technology where he became a Professor in the Civil Engineering Department. These years were very productive in his professional life and were also the years when he helped to form a number of close associates and students who later became part of one of his important legacies. In fact, in 1957 he invited me to join his research group, thus changing the course of my life. In 1961, he left the Illinois Institute of Technology and joined the Catholic University of America in Washington DC. There he had another batch of prominent students and postdoctoral students. In 1963–1964, he was a visiting lecturer at Princeton University and in 1968 he became a Fulbright Scholar. In 1975, he retired from the Catholic University of America and in the subsequent years he became a Visiting Professor at Oakland University in Michigan and at the University of Maryland.

An engineer by vocation, he combined both rigorous scientific approach and a practical sense in the solution of actual engineering problems of significance. As the leader of a research group at the service of industry and the U.S. defense system, he provided solutions for a wide range of problems: stresses and strains in solid propellants, stresses in the structure of missiles, wave propagation problems, ordinance problems, stresses in dams, stresses in nuclear reactors, stresses in soil caused by mining and drilling for oil, and stresses in turbine jet blades. His approach was always based in experimental mechanic techniques. He used all the tools of the trade and continuously sought new techniques if the case under study posed questions that could not be answered by existing methods. Trained as a photoelastician, he introduced the diffused light polarizcope with large field in America so that photoelasticity could be utilized as a practical tool. He developed a number of new techniques in dynamic photoelasticity, and in 3-D photoelasticity. He introduced a number of extremely useful procedures to utilize photoelasticity to solve design problems such as minimizing weight. As we have mentioned before, he developed the theoretical basis for the brittle lacquer technique. He contributed to the development of the Moiré method and wrote a textbook on this subject that is a classical piece in this field. He also worked in the fields of holography, failure of materials, composite materials, and problems involving large deformations. His experiences were presented in four books and in more than 200 papers, reports, and technical notes. One of his most precious legacies has been the training of many outstanding disciples, close collaborators, and students. To mention just a few, W.F. Riley, A.S. Kobayashi, J.W. Dally, I.M. Daniel, K.B. Hofer, R.J. Sanford, V.J. Parks, L. Ferrer, R. Marino, V.J. Lopardo, A.J. Clark, Dick Marloff, myself, and my former doctoral student,

Dr. F.P. Chiang. Through them he has influenced the development and teaching of Experimental Mechanics in the second half of the last century. He pushed all of his students and associates to be active participants in the Society. He was an active participant himself and published a great deal of his work in the different versions of what is known today as *Experimental Mechanics*. He lectured in universities and research institutions throughout the world. He died in March 2000, a month short of his 90th birthday.

Going ahead with what C.E. Taylor called "derivatives," let us start with the close collaborators of Durelli.

Bill Riley. A dear friend and colleague who was a remarkable individual. Can I call him a World War II hero? He was a Lieutenant Colonel in the United States Army Air Force, from 1943–1946. He was a bomber pilot in the skies of Germany and a survivor of this extremely dangerous saga, where the possibility to come back from a mission was low. Being a unassuming person, he did not talk often about this period of his life. Bill was born on March 1, 1925 in Allenport, Pennsylvania where, in 1951, he obtained his Bachelor of Science in Mechanical Engineering at the Carnegie Institute of Technology while attending Max M. Frocht's classes. He received his M.S. in Mechanics under A.J. Durelli at the Illinois Institute of Technology in 1958. He held the following positions at the IIT Research Institute: Research Engineer Armour Research Foundation as it was called at the research institute at that time, 1954–1961, working with A.J. Durelli; Section Manager IIT Research Institute, 1961–1964; and Science Adviser, 1964–1966 where he headed the A.J. Lab. He joined the academic world as a Professor at Iowa State University, Ames, 1966–1978. He became a Distinguished Professor of Engineering, 1978–1988, and Professor Emeritus in 1989. He was an educational consultant at the Bihar Institute of Technology, Sindri, India in 1966 and at the Indian Institute of Technology, Kanpur during the summer of 1970. He was a great educator and a prolific writer. He published, if my accounting is correct, the following books: *Photo Mechanics*, co-authored with A.J. Durelli, *Experimental Stress Analysis*, co-authored with J.A. Dally, and *Mechanics of Materials, Engineering Mechanics, Dynamics*. He was a very active member of our Society and was made a Fellow of the Society. He received the Society's Frocht Award and was made an Honorary member.

Albert S. Kobayashi. Albert was born in Chicago, Illinois in 1924. For his university studies, he returned to Japan, the homeland of his parents. He received his Bachelor's degree at the University of Tokyo in 1947. After graduation, he went to work as an engineer at Konishiroku Photo Industry, Japan. He returned to the U.S. to pursue his graduate studies and got his Master's degree at the University of Washington and a Ph.D. in the Mechanical Engineering Department of IIT. After his graduation, he worked for Illinois Tool Works and in 1956 he joined the Durelli research staff at the Armour Research Foundation. He left IIT to become an Assistant Professor at Washington State University where he also became a Full Professor and, in 1997, became Professor Emeritus. He was rehired on a 40% basis through June 2002 and his funded research continued to June 2005. Albert was a very active participant in the diverse programs in the A.J. group that we have outlined, i.e., the brittle coating work, photoelastic studies, and problems related to dynamic fracture. Similarly to other members of the group, he was a devoted member of SEM. He rose to the presidency of SEM in 1990 and received the following awards: F.G. Tatnall Award in 1973, B.J. Lazan Award in 1981, R.E. Peterson Award in 1983, William M. Murray Lecture Medal in 1983, and M.M. Frocht Award in 1995. He also received the highest award of the Society when he was chosen as an Honorary Member. Albert was also a dedicated teacher and a prolific writer, with the number of his publications exceeding 500. He was the editor and author of the SESA Monograph, *Experimental Techniques in Fracture Mechanics*, Volumes I and II; the editor of the *Manual of Experimental Stress Analysis*, 2nd and 3rd editions; and the editor of the 1st and 2nd editions of the *Handbook of Experimental Mechanics*. His professional activities show remarkable spread and magnitude. He was a consultant (structural analysis and fracture mechanics) to the Boeing Aerospace Company from 1958–1976 and to Mathematical Sciences Northwest from 1962–1982. He was also a consultant to the Air Force Rocket Propellant Laboratory from 1974–1978, to the United Nations Development Program in November 1984, and to the Office of Naval Research from January–June 1986. He was a member of the Committee on Material Response to Ultra-High Loading Rates, the National Material Advisory Board from 1977–1979, the U.S. National Committee on Theoretical and Applied Mechanics from 1991–1995, and the Program Director of the Solid Mechanics Program, National Science Foundation from 1987–1988. He was a Visiting Professor in July 1996 and October 1997 at University de Poitiers, France. From April–June of 2005, he served as an expert in the Solid Mechanics and Structure of Materials Program, National Science Foundation. He was also a member of the National Academy of Engineering.

James W. Dally.   It is a difficult task to express in a few paragraphs Jim's brilliant professional carrier as a teacher, researcher, and engineer. He made fundamental contributions to the developments of experimental methods employed to study dynamic fracture mechanics, and stress wave propagation. He is a prolific writer, having been the author or co-author of about 17 books and a very large number of papers on the subjects that have been mentioned. He worked extensively in literature for graduate and undergraduate education. Jim earned his Bachelor's and Master's degrees in Mechanical Engineering from Carnegie Institute of Technology where he attended Max M. Frocht's classes, in 1951 and 1953 where he earned his Ph.D. in Mechanics from the Illinois Institute of Technology in 1958 where A.J. Durelli was his thesis advisor. In 1961, Dally became Assistant Director of the Mechanics Division of the Armour Research Foundation. When Max Frocht retired in 1964, Dally left the Foundation and joined the Department of Mechanics of IIT and remained there until 1972. In 1972, he became a Professor at the U.S. Air Force Academy at Colorado Springs, Colorado. In 1978, he joined the University of Rhode Island where he was Dean of the Engineering School. In 1984, he became a Professor at the University of Maryland where he remained until his retirement in 1997. He became President of SESA/SEM in 1970 and was made an Honorary Member. More recently, the Society created the J.W. Dally Young Investigator Award. Among the many honors that he received, he was elected to the National Academy of Engineering (1984) and was selected by his peers to receive the Senior Faculty Outstanding Teaching Award in the College of Engineering (1991) and the Distinguished Scholar Teacher Award (1993) at the University of Maryland. He was a member of the University of Maryland team that received the Outstanding Educator Award sponsored by the Boeing Co. (1996), and, more recently, he received an Outstanding Alumni Award (2009) from the Illinois Institute of Technology's Mechanical Engineering Department, the 2012 Daniel C. Drucker Medal from ASME, and the Archie Higdon Distinguished Educator Award from the Mechanics Division of ASEE in 2013.

Following the order of graduation, the next member of the Durelli group is me.

Cesar A. Sciammarella.   In the early 1950s as a Civil Engineering student, I was asked to put a Photoelasticity Lab back together. A main component of the lab was a Favre's Photoelastic setup that provided isoclinics and isochromatics. The setup included a Mach Zhender interferometer that measured the absolute retardations, thus providing the complete solution of the 2-D state of stresses. I put the interferometer back to work and soon enough another professor in the Construction Department asked me to analyze a large concrete structure. The information that I obtained utilizing the interferometer was utilized in building the actual structure. I wrote papers on this work and A.J. Durelli became aware of my work. Then the episode that I have narrated before involving A.J. took place. In early 1957, A.J. asked me if I was willing to join him at IIT to get a Ph.D. I was, at that time, a Professor at the School of Engineering of the University of Buenos Aires and a Professor at the Argentine Army Engineering School. I took my chance and this invitation changed my life.

It is easiest to list the other categories of my professional career and contributions to the field of experimental mechanics by taking sections from my curriculum vita.

Cesar's education includes: a diploma in Civil Engineering, University of Buenos Aires; in 1950 a Ph.D. from the Illinois Institute; in 1960, under thesis advisor A.J. Durelli, the thesis titled, "Experimental and Theoretical Study on Moiré Fringes."

Cesar's academic positions began in 1961 as an Associate Professor, Department of Engineering Mechanics University of Florida at Gainesville and in 1963, Full Professor. In 1967, Cesar became a Professor in the Department of Aerospace, Applied Mechanics, at the Polytechnic Institute of Brooklyn, NY; in 1972, professor and Director of the Experimental Mechanics Laboratory, Illinois Institute of Technology; in 1997, Emeritus Research Professor and Director of the Experimental Laboratory until 2003. From 1991–1998, Cesar was a Non-Resident Professor, Universita degli Studi at Nuoro, Italy. Currently, Cesar is an Adjunct Professor at Northern University of Illinois, Dekalb, IL.

Cesar's Visiting professorships include: 1972 and again in 1976, Polytechnic Institute of Milano, Milano, Italy; University of Cagliari, Cagliari, Italy, 1979; Polytechnic Institute of Lausanne, Lausanne, Switzerland, 1979; University of Poitiers, Poitiers, France, 1980; and a Professor at the Polytechnic Institute of Bari, Bari, Italy, 1992, 1994, 1998, and from 2003–2008.

In 1961, Cesar's research included analysis of optical signals connected with displacements and derivatives of displacements, continuation of the Ph.D. work on the Moiré method. In 1965, generalization of the concept of fringe order introduced in Photoelasticity as an integer number, utilizing the Fourier transform extended to all real numbers. Concrete results of this paper, (SEM Spring meeting 1968), first paper in the world literature of fringe pattern analysis that presented and illustrated computer processing of fringe patterns. A U.S. patent was issued for a system to scan photographic negatives, convert gray levels to numbers, and get perforated tapes as input of IBM main frame computers. What factors control the possible resolution that can be achieved in space and frequency utilizing advanced algorithms and computers? In 2000, a paper based on the Heisenberg principle provided an answer to this question. From 2015–2018 a mathematically based self-consistent approach covered a wide spectrum of scales, macro, micro, nanometer, and sub-nanometer scales.

The applications area for Cesar included different developments as applied to Moiré patterns, speckle patterns, and holographic patterns in 2-D and 3-D applications. Two valuable cross-matching concepts evolved in these applications: the Holographic-Moiré, application of the developments in Moiré fringe patterns to holography, and vice versa the Moiré-holography, utilizing of developments in holography to Moiré patterns. This idea implies applying holographic concepts developed in the range of the frequency of light to the frequency of a selected grating pitch.

Application of optical techniques and other methods of experimental mechanics to important scientific and technical problems was also a focus such as in the areas of: fracture mechanics of metals both linear and nonlinear; 2-D and 3-D, detailed analysis of the crack tips elastic, plastic, and pre-fracture; an extensive study on the fracture of high strength steel alloys vessels under internal pressure; fatigue/fracture of metals using gamma-ray's spectroscopy; composite materials: fracture of single plies of glass-epoxy under dynamic loads in the range of strain rates of 0.01 1/min to 100,000 1/min; particulate composites: study of the brittleness of metal matrix reinforced with hard particles; extensive study of the damage mechanism of particulate composites under repeated tensile loading with high-resolution microscopy and modeling with finite elements; and rolling fatigue, wear, and fracture of railroad wheels.

Contributions in bioengineering included the study of the vertebral column mechanical properties and effects of different types of reinforcements introduced by surgery of the vertebral column; the study of knee-cap prosthesis contact stress damage; and extraction of displacements and strains of the heart from MRI images.

Contribution to the field of sub-wave length optics included: in 2005, going beyond the Rayleigh limit, holographic optical detection of information at the sub-wavelength level; and, in 2010, light generation at the nano-scale and holographic interferometry at the nano-scale.

Applications to metrology included the replacement of mechanical sensors of CMM machines with optical reading heads and high-accuracy techniques to measure shape of 3-D bodies.

Cesar has three US patents. His other works/projects of note include the following.

- United Nations. United Nations Development Program. Planning, setting up, training personnel, and guiding the development period of an Experimental Mechanics Laboratory at the Structural Engineering Research Center of the Council of Scientific Research and Industrial Research of the Indian Government, Madras, India, 1974–1980.

- Participation in important engineering projects. In the 1960s Saturn V/Apollo project and in the 1970s worked on a critical component of the space shuttle. In 1983, during a sabbatical year, head of the structural designing group of a 50 m steel sphere, a containment building of a 500 Mw nuclear reactor in the Argentine republic.

- Awards received by Cesar include: SEM Fellow, Hetenyi, Lazan, Frocht, Theocaris awards and the William M. Murray Medal. In 2013, Honorary Membership. In 2011, the British Society for Strain Measurement, Fylde Electronic Price for best paper published in *Strain* in 2010. In 2013, Polytechnic of Bari Life Achievement Award. In 2014, Life Achievement Award, International Conference on Computational and Experimental Engineering and Sciences, Changwon, Korea. Lifetime member of the American Society of Mechanical Engineers and Honorary Member of the Italian Association of Stress Analysis. European Society for Experimental Mechanics (EURSEM), 2016 Lifetime Award.

Isaac Daniel.   Isaac was born in the 1930s in Thessanoliki Greece and as a young child escaped death at a concentration camp when the Germans occupied Greece. An Italian consul helped the Daniel family to get papers to seek refuge in Italy, where her mother was born a member of the family of the famous painter Modigliani.

Isaac inherited A.J. Durelli's Laboratory at IITRI. He became Manager of the Experimental Stress Analysis Section, Science Advisor, Materials and Manufacturing Technology Division of IITRI in 1966.

Isaac received his education at IIT—a B.S. in 1957, an M.S. in 1959, and a Ph.D. in 1964, all in the area of Civil Engineering. **Academic Positions**. 1986–present Professor of Civil and Environmental Engineering and Mechanical Engineering, Northwestern University; 1996–2009 Director, Theoretical and Applied Mechanics Program of Northwestern University; 1997–present Director, Center for Intelligent Processing of Composites.

Isaac's research proved him to be an accomplished and outstanding member of the solids mechanics community in its theoretical aspects as well as a remarkable experimentalist. Isaac's research encompasses many areas of mechanics and materials with particular emphasis on experimental mechanics techniques utilized to create a consistent theory of composite materials. He has covered all aspects of the mechanics of composites including: processing, characterization, and modeling of polymer, ceramic, and metal matrix composites; nanocomposites and hybrid nano/microcomposites; fracture and damage mechanics; nondestructive evaluation; and life prediction. He has introduced test methods for the characterization of composites and has developed a new failure theory that is universally accepted and utilized in the design of composites. He co-authored a book in composites that is a classical book both in education and in engineering practice. He has published hundreds of monographs on the topic of composites.

Isaac's awards and recognitions include: Editorial Board of the *Journal of Composite Materials*; Editorial Board of *Composites, Part A*; Fellow of the American Academy of Mechanics; Fellow, Society for Experimental Mechanics (1981); William M. Murray Medal and Lecture; Honorary Member (2007); Technical Editor; Executive Board; Fracture Committee; Composites Committee; Monographs Committee; Committee on Fellows; International Advisory Board (1980–2012); Associate Editor of *Journal of Applied Mechanics* (1993–1999); Member of the European Academy of Sciences (EAS), 2009; Member of the Editorial Boards of *Composites A, Journal of Composite Materials, and Strain*; Foreign member of Russian Academy of Engineering, 2012.

Vincent Parks.   Another important member of the Durelli team was Vincent J. Parks. He was personally very close to A.J. and his right-hand in many of his activities. He followed A.J. to the Mechanics Division of the Engineering School of the Catholic University of America in 1961, where he became an Assistant Professor and then an Associate Professor. He was the co-author of many of the papers that A.J. published during his tenure at the Catholic University and co-author of A.J.'s famous book on Moiré.

We come now to the second generation of the IIT-IITRI branch, the students that actively continued the connection of professors with SEM.

Arun Shukla.   Simon Ostrach Professor, Mechanical, Industrial, and Systems Engineering, University of Rhode Island.

Arun's educational background includes a B.SD. Indian Institute of Technology, Kanpur 1976, Ms. and Ph.D. from the University of Maryland 1981, thesis advisor, J.W. Dally.

His academic positions include the University of Rhode Island: 1981–1984 Assistant Professor; 1984–1988 Associate Professor. 1988–present Full Professor; 2000 Chair Professor; 1997–2000 Chairman of the Department of Mechanical Engineering and Applied Mechanics; 2002–2003 Interim Dean of the College of Engineering.

He was also a visiting professor in the Department of Engineering and Applied Science Aerospace at the California Institute of Technology, 2011, Texas A & M 1944–1945, Department of Mechanical Engineering, Indian Institute of Technology, Kanpur, sabbatical leave 1987–1988.

Arun's professional memberships and activities include: Member Executive Committee, Applied Mechanics Division, ASME (2012–2017); President Society for Experimental Mechanics (2002–2003); President Elect Society for *Experimental Mechanics* (2001–2002); Vice President, SEM (2000–2001); Technical Editor, International Journal, *Experimental Mechanics* (1997–2000); Associate Technical Editor; *Experimental Mechanics* (1987–1996); Associate Technical Editor International Journal, *Lasers and Optics in Engineering* (1997–2012); Member Editorial Board, *Strain* (2006–Present); Member Editorial Advisory Board, Key En-

gineering Materials, *Trans Tech Publications* (1997–Present);
Member Editorial Board, International Journal, *Lasers and Optics in Engineering,* (1993–1996);
Member, Executive Board SEM (1994–1996); Chairman, Technical Divisions Council, SEM, (1995–1996); Chairman, Fellow's Committee, SEM (1996–1997, 2011–2012);
Chairman Fracture Mechanics Committee, ASME, 1992–1995; Chairman, Fracture and Fatigue Division, SEM, 1992–1993; Member, Honors Committee, SEM, 1991–1994, 2003–2006;
Member, ASEE, AAM, ASME, ISTAM, and SEM.

He continued his Ph.D. thesis work on dynamic crack propagation utilizing photoelasticity. Another area of his work is the wave propagation in granular, layered, and porous media. He has worked in the dynamic fracture of composite materials, dynamic characterization of advanced polymers, damage and damage growth in particulate composites, and impact damage in concrete and granite. He developed techniques of ultra-high-speed imaging. He worked on experimental and analytical evaluation of dynamic fracture in graded multifunctional materials. He has made very valuable contributions to blast loading in different environmental conditions.

Arun has published more than 200 papers in his area of interest and has more than 3,500 citations. He published one book as a co-author of J.W. Dally, *Experimental Solid Mechanics* (2015) and is the author of *Practical Fracture Mechanics Design* (2005, Marcel Dekker, New York).

His awards and honors include: Elected to European Academy of Sciences and Arts, 2011; Fellow of the American Academy of Mechanics, 2001; Fellow of the American Society of Mechanical Engineers, 1996; Fellow of the Society for Experimental Mechanics (SEM), 1993; Murray Medal, SEM 2011; Taylor Award for Technical Excellence in Optical Stress Analysis, SEM 2012; Tatnall Award for Long and Distinguished Service to SEM 2012; College of Engineering Faculty Excellence Award, 2012; University of Rhode Island Outstanding Contribution to Research Award, 2001; Distinguished Alumnus Award, Indian Institute of Technology, Kanpur, 2009; B. J. Lazan Award for Outstanding Technical Contributions to Experimental Mechanics, 2002; *Educator of the Year—M.M. Frocht Award*, 2001; University of Rhode Island Research Achievement Award (2001); Vincent E. and Estelle Murphy Faculty Excellence Award, College of Engineering, URI, 1998; ASTM Outstanding Paper Award J. of Testing and Evaluation, 1998; The University of Rhode Island's Scholarly Excellence Award, 1995; Albert E. Carlotti Faculty Excellence Award, College of Engineering, URI, 1990.

Next is the third generation of A.J. Durelli that continued the SEM affiliation.

Fu-Pen Chiang. Fu-Pen is the Distinguished Professor and Chair of Mechanical Engineering of the State of New York University at Stony Brook and is the Director of the Laboratory for Experimental Mechanics Research, State University of New York at Stony Brook.

Fu-Pen got his B.S. degree in 1957 from the National Taiwan University Department of Civil Engineering, his M.S. degree in 1963, and a Ph.D. in 1966 from the Mechanics Department of the University of Florida under my guidance. He continued his collaboration with me in

1967 at the Polytechnic of Brooklyn. During the period of 1967–1968 he was a Post-Doctoral Fellow with A.J. Durelli at the Catholic University of America.

Fu-Pen's visiting professorships include: from 1973–1974 the Swiss Federal Institute of Technology at Lausanne; 1980–1981 Senior Visiting Fellow at Cambridge England; and 1990–1991, Visiting Professor National Taiwan University, Taiwan.

His research emphasis is on the development of optical and other experimental mechanics techniques and their applications to stress analysis (including NDE), fracture, and fatigue of engineering and biological materials and structures

Fu-Pen is a prolific writer that has covered vast areas of research in theoretical and experimental mechanics. He has published 56 papers in the theory and application of the Moiré Method; 6 papers on Photoelasticity; 62 papers on the subject of Coherent Light Speckle Methods and applications; 82 papers in White Light Speckle Photography and applications; 12 papers in Holographic and Holographic Interferometry; 36 papers in Optical Metrology and on Non-Destructive Evaluation; 154 papers in the area of Fracture, Fatigue, and Failure of Solids; 20 publications in the area of Biomechanics; and 15 monographs and invited review article.

Fu-Pen's honors and awards include: Lifetime Achievement Medal (2012) International Conference on Computational & Experimental Engineering and Sciences; Fellow of American Society of Mechanical Engineers (2011); Fellow of SPIE—The International Society of Optics and Photonics (2011); SEM, 1983 Fellow; 1993 Lazan award; 2009 Frocht award; 2015 Theocaris award; 2018 Murray Lecture; Fellow of Optical Society of America (1988); and Lifetime Achievement Award (2016) from the International Conference of Experimental Mechanics.

John Gilbert.   When I became a Professor at IIT in 1971, I made sure that John, at that time just finishing his Master's degree under my guidance, would be accepted as a graduate student and future candidate for a Ph.D. At that time in New York, John had extracurricular activities such as guitar playing and was the band leader of a group of young musicians. A colleague of mine from whom he was taking classes told him that he would never amount to much in his life. From that moment, John took as a task in his life to prove him wrong…the rest is history.

John's education included a Ph.D. in Mechanics from the Illinois Institute of Technology, 1975; M.S., Applied Mechanics, Polytechnic Institute of Brooklyn, 1973; B.S., Aerospace Engineering, Polytechnic Institute of Brooklyn, 1971.

His academic positions from 1975–1985 included: Assistant Professor (1975–1981), associate professor (1981–1985), University of Wisconsin Milwaukee, Department of Civil Engineering; 1985–2013, Professor; 2013 Professor Emeritus, at the University of Alabama. Simultaneously with these positions he was: 1989–2013, Adjunct Professor, University of Alabama at Birmingham, Materials Science; 1988–2013, Adjunct Professor, University of Alabama in Tuscaloosa, Materials Science; 1988–1990, director of Civil Engineering program, University of Alabama in Huntsville, Department of Mechanical Engineering.

As a Professor he continued his research work in the field of his Ph.D. thesis of holography, holographic interferometry, and holographic Moiré. He published many papers in the utilization of fiber optics to get holograms in not directly accessible locations and extended this work to speckle interferometry and speckle interferometry. He also worked on the application of these developments to non-destructive evaluation and in the sensing of ultrasound waves. He introduced his own brand of radial interferometry utilizing radial lenses and worked in the characterization and fabrication of radial lenses, and the generation of holographic interferometry patterns. He applied this technology to the inspection of pipes and cavities. He also worked in the development of fiber-optic accelerometers. More recently, he concentrated in the development of structural components made out of cementitious steel-reinforced components to create high temperature-resistant structures for aerospace applications.

John co-authored with me a number of papers on a seminal subject of holographic Moiré and application to the measurement of 3-D displacement of transparent bodies. He continued to publish papers on this topic. He published many papers on panoramic holographic interferometry and its application to pipes in the aerospace industry. He published papers on the development of high-strength cementitious materials and their possible application to aerospace industries and numerous papers on optical metrology. John holds three US patents.

John's awards and honors include: A Fellowship Award from the American Society for Nondestructive Testing for the development of a thermal acoustic photonic system for nondestructive evaluation and awards from NASA and the AIAA for pioneering work in the areas of radial metrology, fluid flow visualization, and panoramic imaging in space. Portions of his work were declared as "milestones in optics" by SPIE and he was recognized by ASM for the evolution of high compliance ratio cementitious composites. His academic awards include the UAH Alumni Association's Distinguished Faculty Award, the Foundation's Distinguished Teaching Award, the College of Engineering's Outstanding Faculty Member Award, the UAH Student Government Outstanding Faculty Award, and the UAH Outstanding Student Group Advisor Award. His professional awards include ASCE's Engineer of The Year Award and HAT's Professional of The Year Award. Dr. Gilbert is a fellow of the Society for Experimental Mechanics and is, or has been, a member of the British Society for Strain Measurement, the Society of Photo-Optical Instrumentation Engineers, the American Academy of Mechanics, the American Society for Engineering Education, the American Society of Civil Engineers, the Huntsville Electro-Optical Section and Working Group, Sigma Xi, and the Order of the Engineer. He was twice designated as an International Man of the Year by the International Biographical Center and declared Man of the Year by the American Biographical Institute. Citations include Continental Who's Who Registry of National Business Leaders, Empire Who's Who Among Executives and Professionals in Engineering, Who's Who in Technology Today, Who's Who in American Education, Who's Who in Science and Engineering, 2000 Outstanding Intellectuals of the 20th and 21st Centuries, 2000 Outstanding Scholars of the 20th and 21st Centuries, and Who's Who in the World. He has supervised 14 Ph.D. dissertations, as well as 39 M.S. and 8 B.S. theses to completion and is currently supervising several graduate students. His major research projects include "Strategically Tuned Absolutely Resilient Structures" (U.S. Army) and "Creating Resilient Structures with Polymer Coatings" (U.S. Department of Commerce).

CHAPTER 4

The Taylor Legacy/Series–The University of Illinois at Urbana Connection

I have gone into detail regarding the IIT-ITRI connection that started early in SEM's development. We will now shift our attention to C.E. Taylor's outstanding contributions to experimental mechanics and his "derivatives." He was the source of many of the most prominent members of SESA/SEM.

C.E. Taylor (1924–2017). Chuck's education included a B.S. 1946, Mechanical Engineering, Purdue University, West Lafayette, Indiana; M.S. 1948, Engineering Mechanics, Purdue University Engineering Mechanics, Purdue University; Ph.D. 1953, Theoretical and Applied Mechanics, University of Illinois at Urbana.

His academic positions began 1946–1948 as an Instructor at Purdue University; 1948–1951, Instructor, Department of Theoretical and Applied Mechanics University of Illinois, Urbana; 1951–1952, Assistant Professor, Applied and Theoretical Mechanics, Urbana. In the next two years, 1952–1954, he left academia to become a structural research engineer at the David Taylor Model Basin.

In 1954, he returned as an Assistant Professor at Urbana, in 1955 he became an Associate Professor, was made Full Professor in 1957, and in 1981, he became Professor Emeritus. In 1981, he became Professor, Engineering Sciences, University of Florida at Gainesville. In 1991, he became Professor Emeritus.

Chuck's visiting professorships included: 1966, Department of Mechanical Engineering, Bihar Institute of Technology, India; 1967, Visiting Professor, Department of Civil Engineering, University of Colorado, Boulder; 1968 Professor Mechanical Engineering University of California, Berkeley, California; 1969, Visiting Professor, Department of Mechanical Engineering India Institute of Technology, Kanpur, India; and 1972, Visiting Professor, Technische Hogeschool, Delft, Netherlands.

Chuck's professional activities included: the Society for Experimental Stress Analysis (SESA) (now SEM), President, 1966–1967, many committee chairmanships, Historian and Chairman of Past Presidents Committee. American Society of Mechanical Engineers (ASME), Associate Editor 1971–1972, *Journal of Applied Mechanics*, several committees. Society for Engineering Science (SES), President, 1978, several committee chairmanships. American Academy of Mechanics (AAM), President 1993–1994, several committees.

His honors and awards included: 1968, SEM Past President Award; 1969, SEM M.M. Frocht Award; 1970–1973, M. Hetenyi Awards; 1974 William M. Murray Lectureship; 1975, SEM Fellow; 1979, Elected to National Academy of Engineering; 1980 American Association for the Advancement of Science Fellow; 1980 Commanders Award for Civilian Service, Department of the Army; 1983 SEM F. G. Tatnall Award, Honorary Membership; 1991 ASME Life Fellow; 1998 AAM Award for Distinguished Service to Theoretical and Applied Mechanics; 2000 SEM creation of the C.E. Taylor award, first recipient; 2003 Honorary Doctor of Engineering, Purdue University.

He contributed both to Theoretical and Applied Mechanics in the area of shells and structures. He published work in the area of Photoelasticity and holographic interferometry applied to Photoelasticity. He contributed to dynamic recording of holographic patterns using pulsed lasers. He also worked on shearing holographic interferometry to measure displacement derivatives and also in holographic-Moiré interferometry and in diverse aspects of displacement recording using holographic interferometry. He contributed to the dynamics of crack propagation in the field of Fracture Mechanics. He also published papers in the area of speckle interferometry and in the application of fiber optics to the measurement of strains and in the computerization of fringe pattern analysis.

He was the co-author of approximately 50 papers in the previously indicated fields. He also was the co-author of a book *Mechanical Behavior of Solids* and contributor to chapters in the *SEM Manual on Experimental Stress Analysis*.

The lists of his Ph.D. students that have been active members of SESA/SEM and outstanding teachers and researchers in the fields of Theoretical and Experimental Mechanics is impressive. This list includes: Daniel Post 1957, R.E. Rowlands 1967, D.C. Holloway 1971, W.F. Ranson 1971, Y.Y. Hung 1973, J. F. Doyle 1977, Y.J. Chao 1981, William M. Fourney, Michael Sutton 1981.

Let us go, over the "derivatives" of C.E. Taylor in order of graduation. The first outstanding student of Taylor's is Daniel Post. Dan, has achievements in many different fields of Experimental Mechanics always characterized by his extraordinary skills as an experimentalist that made his work unique in many different areas.

Daniel Post.   Dan was born in Brooklyn, NY in 1929.

Dan's education included: 1947–1948 Pratt Institute New York; 1948–1950 B.S. University of Illinois at Urbana; 1948–1950 M.S., 1955–1957 Ph.D. in Theoretical and Applied Mechanics, advisor C.E. Taylor.

His academic positions included: 1955–1956 Instructor, Theoretical and Applied Mechanics, University of Illinois, Urbana; 1958–1960 Adjunct Associate Professor; 1964–1966 Associate Professor Rensselaer Polytechnic Institute, Connecticut; 1978–1991 Professor; 1991 Emeritus Professor, Engineering Science and Mechanics Virginia Polytechnic, Blacksburg, VA.

His visiting professorships included: 1965 University of Columbia, New York; 1965–1966, Department of Mechanical Engineering Sheffield, England.

He was a consultant to may government institutions, such as the U. S. Naval Research Laboratory and private companies particularly Photoelastic Inc., a Division of Vishay Intertechnology.

Chuck's honors and awards included: being an active member of SESA/SEM, participating in many Divisions of the Society. In 1971, Fellow, 1977 Murray Award, 1999 Honorary Member, 1985 Royal Society Award, London for collaboration research at the University of Oxford.

Beginning in 1951 and extending to 2006, Dan focused on the development and application of practical techniques and tools for experimental solid mechanics. In 1953–1963, when Photoelasticity was the dominant method of full-field stress analysis, he developed Photoelastic Fringe Multiplication and other special techniques in Photoelasticity including dynamic crack propagation. He also did important work in Scattered Light Photoelasticity and Oblique Incidence Photoelasticity. In 1963, Post developed a unique method to manufacture foil-type electric strain gages. The method is used for special gages by Vishay Intertechnology, and it is used extensively for the production of precision resistors by the same company. During subsequent years, Dan studied Moiré techniques for displacement and strain measurements, and published several incremental techniques for Moiré fringe multiplication. Multiplications by factors up to 60 were demonstrated. In 1979, Post introduced a scheme of Moiré interferometry utilizing white light. He also introduced virtual gratings, a concept taken from optical holography. His achievements in this area are remarkable showing an extraordinary skill in implementing theoretical developments into actual instrumentation.

Dan has three U.S. patents.

Dan has a large number of publications in his field of work as well as a book, *High Sensitivity Moiré* (Springer-Verlag 1994).

R.E. Rowlands. Robert (Bob) Rowland's education included: B.S.c 1959, University of British Columbia (Vancouver, Canada); M.S. 1964, University of Illinois Urbana-Champaign; Ph.D. 1967, University of Illinois Urbana-Champaign.

His academic positions include: Professor Mechanical Engineering, University of Wisconsin at Madison. Upon graduation he joined the group headed by Isaac Daniel at IIT Research Institute in Chicago where he worked for eight years in research and development in experimental mechanics, stress analysis, and composite materials

His research activities are focused in using infrared thermography (thermoelastic stress analysis), Doppler Interferometry, microscopy, and Moiré methods for stress and fracture analysis of composite structures, fiber-reinforced concrete, bolted joints, metal plate connectors in wood structures, and cellulosic composites. For six years he served as director of the Wisconsin Structures and Materials Testing Laboratory.

He is a fellow of the American Society of Mechanical Engineers and the SEM. Additionally, he belongs to the American Academy of Mechanics and the American Academy of Composites and is a registered professional engineer. He is an active member of SEM, a Fellow of SEM, and received the Felix Zandman Award in 2014 in recognition of the use of birefringent coating for stress analysis of bolted wood joint/connections.

He has published numerous papers in his areas of interest.

David C. Holloway.   David C. Holloway is a member of the of Mechanical Engineering at the University of Maryland and a co-author of papers with C.E. Taylor working in the field of recordings of dynamic events. He retired in 2004 after 30 years of service after devoting his efforts to the education field.

William H. Ranson.   Currently Distinguished Professor Emeritus at the University of South Carolina, Columbia, SC. How is it possible to characterize Bill Ranson in a few words? Bill has a brilliant mind and was a prime mover of seminal theoretical and experimental endeavors in the field of solid mechanics.

Bill coauthored papers in digital imaging techniques to measure surface displacement components in laser speckle techniques. He also co-authored papers in the area of photoelasticity and the mathematical modeling of mechanical systems. Bill has been a member of SEM since 1983.

Y.Y. Hung.   Y.Y. Hung is one of the most highly successful students of C.E. Taylor that has had an important impact in the field of application of optical methods of Experimental Mechanics.

Y.Y.'s education included: M.S. and Ph.D. in 1973 University Illinois at Urbana, Department of Applied Mechanics.

After graduation from the University of Illinois he became an Associate Research Engineer, IIT Research Institute, working in the research group headed by Isaac Daniel.

His academic positions include: Assistant Professor at the Mechanical Engineering Department Oakland University, Department of Engineering and Computer Sciences, where he spent a great deal of his academic life becoming a Full Professor and finally retiring as Emeritus Professor in 2002 to join the Engineering and Engineering Management Department at the City University of Hong Kong where he had the position of Full Professor and Department Head, finally retiring in 2008 as Emeritus Professor.

He continued during his career working in the topic of his doctoral thesis. He coined the word *Shearography*. Shearography is a laser-based technique for full-field, non-contacting measurement of surface deformation (displacement or strain). It was developed to overcome several limitations of holography by eliminating the reference beam. It does not require special vibration isolation, hence, it is a practical tool that can be used in a field/factory environment. Shearography has a received industrial acceptance, in particular, for non-destructive testing. In non-destructive testing, shearography reveals defects in an object by identifying defect-induced deformation anomalies. Other applications of shearography include strain measurement, material characterization, residual stress evaluation, leak detection, vibration studies, and 3-D shape measurement. Besides this very successful technology that he introduced and developed, as a professor in the area of Michigan that was connected to the automotive industry, he worked in the applications of diverse optical methods to the automotive industry.

An active member of SEM he received the Lazan award in 1997.

He has a large number of publications in the field of Shearography both in the theory and its applications in prestigious journals and many other topics of applied optics to Experimental Mechanics.

James F. Doyle.   Jim was also an active member of SESA and has been a Professor at Purdue in the College of Engineering since 1977.

Jim's education includes: Dipl. Eng., Dublin Institutes of Technology, Ireland, 1972; M.Sc., University of Saskatchewan, Canada, 1974; Ph.D., University of Illinois, 1977.

Jim's main research area is Experimental Mechanics with an emphasis on the development of a new methodology for analyzing impact and wave propagation in complicated structures with the goal of being able to extract the complete description of the wave and the dynamic system from limited experimental data. Special emphasis is placed on solving *inverse problems* by integrating experimental methods with computations methods (primarily Finite Element-based methods).

He has numerous journal papers in his area of interest and is an author of the following books: *Wave Propagation in Structures*, Springer-Verlag, New York, 1989, 2/E 1997; *Static and Dynamic Analysis of Structures*, Kluwer, The Netherlands, 1991; *Nonlinear Analysis of Thin-Walled Structures: Statics, Dynamics, and Stability*, Springer-Verlag, New York, 2001; *Modern Experimental Stress Analysis: Completing the Solution of Partially Specified Problems*, Wiley & Sons, UK, 2004; *Guided Exploration on the Mechanics of Solids & Structures: Strategies for Solving Unfamiliar Problems*, Cambridge University Press, UK, 2009.

Jim is an SEM Fellow and has been honored with the Frocht and Hetenyi awards.

Yuh J. (Bill) Chao.   Yuh J. (Bill) Chao is an Endowed Professor at University of South Carolina.

Bill participated in the department of the image correlation work and published many papers on the field of fracture mechanics and material's properties. More recently, he co-authored papers on the subject of nano scale and in the field of biomechanics as well as in the generation of electrical energy in solar cells. Bill has continued his work in material sciences and fracture mechanics.

Bill's honors and awards include: SEM 1996 Peterson award; 2000 Fellow; 2003 Hetenyi Award; 2006 Lazan Award; SEM 2006 Fellow; 2016–2018 SEM Executive Board Member; Research Fellow Award, Alexander von Humboldt Foundation, Germany, 1991.

William (Bill) Fourney.   We have already connected SEM activities with the University of Maryland through J.W. Dally. This tradition was continued by William Fourney. The Fourney brothers, Bill and his brother Mike, are iconic figures in SEM, active members in many different capacities within the Society and contributors to the current Society development and very close friends of C.E. Taylor.

Bill's current academic position is: Keystone Professor, Department of Mechanical Engineering, Department of Aerospace Engineering, and Associate Dean.

Bill has published approximately 250 articles describing the results of his research in dynamic fracture and crack arrest, including approximately 50 reports to the sponsors of his research. His research has been sponsored by NSF, AFOSR, ONR, NRC, DOE, NSWC, TARDEC, ARL, Oak Ridge National Labs, Los Alamos National Labs, and the US Bureau of Mines. He has served as a consultant for Gillett Research Labs, Versar, Allegheny Ballistics Lab, Harry Diamond Labs, Los Alamos National Labs, and SAIC. He is active in the SEM and the International Society for Explosives Engineering (ISEE). Within ASTM he served as Chairman of the Crack Arrest Technical Committee. Within ISRM he was President of the Commission of Fragmentation by Blasting and is North American Editor of *International Journal on Fragmentation and Blasting*.

Bill's honors and awards include: SEM Fellow; 2000 F.G. Tatnall Award; 2015 Frocht Award; 2016 C.E. Taylor Award; AIAA Associate Fellow; Fellow ASME; University of Maryland President's Medal 2006; Member of Distinguished Alumni, University of Illinois-TAM Department.

Michael Sutton.   Carrying on the C.E. Taylor Legacy/Series in South Carolina is Michael (Mike) Sutton. Mike is without a doubt one of the more successful members of the Experimental Mechanics community. Mike has been the leader of a remarkable group of people and a main contributor to the currently most used software system in Experimental Mechanics known worldwide as the area of Digital Image Correlation (DIC).

Mike's academic position started in 1982 where he joined the Faculty of the Department of Mechanical Engineering of South Carolina University and is currently a Distinguished Professor.

Mike's education includes: B.S. and M.S. from the University of Sothern Illinois, and as mentioned before, Mike received his Ph.D. in 1981 in the University of Illinois at Urbana under the guidance of C.E. Taylor.

Mike has a vast array of research areas of his interest. Coherent and incoherent optics applications, digital image processing, applications of integral methods and experimental mechanics, plastic fracture mechanics, and finite element modeling of cracked bodies.

A prolific writer, Mike has published more than 400 papers and monographs in his field of interest. Furthermore, the citations of his work run into the 1000s. He has published a book, *Image Correlation for Shape, Motion and Deformation Measurements*, Springer, 2009. Recently, he was the co-editor with Philip Reu of the *Proceedings of the Annual International DIC Society*

Conference and *SEM Fall Conference* organized by the Society for Experimental Mechanics and Sandia National Laboratories held in Philadelphia, PA, 2016.

Mike's honors and awards include: SEM Fellow 2000; Murray Medal 2013; 2017 Tatnall Award; Fellow ASME 2004, Doctoris Honoris Causa from Ecole' Polytechnic in Cachan-Paris, France 2011.

Having outlined the curriculum vitae of the outstanding members of SEA/SEM that are direct "derivatives" of Chuck, we can go to the next derivatives that have kept alive the tradition that he started.

Peter G. Ifju.   Mechanical Engineering, University of Florida at Gainesville.

Peter's education includes: 1986 Virginia Polytechnic Institute, B.S. Civil Engineering; 1989 M.S. in Engineering Science and Mechanics; 1992 Ph.D., thesis advisor Daniel Post.

Peter is currently a Professor and Associate Chair of the Department of Mechanical Engineering at the University of Florida, Gainesville.

His research work focuses on Experimental Mechanics, with emphasis in the field of Moiré methodology, unmanned aircraft system research and non-destructive evaluation for quality control.

He has published extensively in his area of research interest Moiré methodology and application to solid mechanics, photoelastic luminescent coatings, flight control of aeroelastic fixed wing micro aerial vehicles, assessment of controllability of micro air vehicles, and measurement of residual stresses. He is co-author, with Dan Post, of the book *High Sensitivity Moiré: Experimental Analysis for Mechanics and Materials*.

Peter's honors and awards include: SEM Fellow; 2011–2012 SEM President.

Bongtae Han.   University of Maryland, Department of Mechanical Engineering

Bongtae's education includes a Ph.D. 1991 Virginia Tech, thesis advisor Daniel Post.

He is currently the Keystone Professor at the University of Maryland, Department of Mechanical Engineering.

His research work includes: mechanical design of photonics and microelectronics packaging for optimum reliability; Physics of Failure (PoF) of advanced semiconductor packaging; reliability assessment and life assessment of automotive electronics; mechanical characterization of electronic packaging materials; and experimental micro- and nano-mechanics (optical methods and methodologies.

Bongtae's honor and Awards include: IBM Excellence Award for Outstanding Technical Achievements, 1994; Brewer Award for Outstanding Experimental Stress Analyst, SEM, 2001; Gold Award (the best paper in the Analysis and Simulation session) - The 1st Samsung Technical Conference, November 9–12, 2004; 2004 Best Paper Award - IEEE Transactions on Components and Packaging Technologies, 2005; Fellow - Society for Experimental Mechanics (SEM), 2006; 2004 Associate Editor of the Year Award - *ASME Journal of Electronic Packaging*, 2006; Fellow - American Society for Mechanical Engineers (ASME), 2007; 2015 Best Paper

Award, the 16th International Conference on Electronic Packaging Technology (ICEPT 2015); 2016 Mechanics Award, American Society of Mechanical Engineering, ASME Electronic and Photonic Packaging Division.

CHAPTER 5

Other Significant Contributors to SEM

We have traced the early history of SESA/SEM to the East Coast of the U.S., but there are also important contributions connected to the West Coast and to the California Institute of Technology. This tradition is connected to an outstanding figure in the U.S. scientific and technological community, Wolfgang Knauss.

Wolfgang Knauss. Knauss grew up during World War II in Siegen Germany and is the son of a Methodist Pastor. After the war, through the acquaintance of a Methodist Pastor from Pasadena, CA, Frank Williams, the Knauss family settled in Pasadena where Knauss attended Pasadena City College.

Wolfgang's education includes: B.S., M.S., and Ph.D., Caltech in 1958, 1959, and 1963, respectively. He did his Ph.D. work under the guidance of Max L. Williams, one of the fathers of Fracture Mechanics, on the rupture of viscoelastic materials. This event provided the fundamental endeavor of his life.

His academic positions included: Research Fellow in Aeronautics, 1963–1965; Assistant Professor, 1965–1969; Associate Professor, 1969–1978; Professor, 1978–1982; Professor of Aeronautics and Applied Mechanics, 1982–2001; von Karman Professor, 2001–2004; von Karman Professor Emeritus, 2004.

The topics of his research work are diverse and include basic theoretical aspects as well as technological applications: application of scanning tunneling microscopy to problems of interfacial strength design in composite structures; constitutive behavior of matrix materials for high temperature composites; fracture behavior of non-linearly viscoelastic solids related to adhesive bonding in solid propellant rockets (shuttle booster); fatigue of thermoplastic matrix materials for composites; time-dependent buckling of structures made of fiber-composites failure of and crack propagation in (particulate) composites incorporating microdamage in high deformation gradients; adhesion and interfacial fracture mechanics; and geometry-induced failure of composite structures for future aircraft.

His publications exceed 500 and cover the field of the mentioned research topics. Published with M.L. Williams *Dynamic Fracture*, Dordrecht, Boston, Nijhoff, 1985; Editor with A.J. Rosakis, *Nonlinear Fracture—Recent Advances*, Kluwer 1990; Editor with Richard Schapery,

Failure at and Near Interfaces, PN, 1996; *Recent Advances in Fracture Mechanics: Honoring Mel and Max Williams*, Kluwer 1998.

He has received the following honors and awards: Woodrow Wilson Foundation Fellowship; the National Aeronautics and Space Administration Fellowship; Tau Beta Pi Engineering Honorary Society; Elected Fellow, Institute for the Advancement of Engineering, 1971; National Academy of Sciences Lecturer to the U.S.S.R., 1977; Elected Fellow, National Academy of Mechanics, 1980; Senior U.S. Scientist Award by the Alexander von Humboldt Stiftung, 1986/87; Fellow, Society for Experimental Mechanics, 1987; Murray Medal, Society for Experimental Mechanics, 1995; Fellow, American Society of Mechanical Engineers, 1996; Corresponding Member of the International (Russian) Academy of Engineering, 1996; 2010 Timoshenko Medal from the American Society of Mechanical Engineers "for fundamental contributions to the mechanics of fracture, covering mixed-mode fracture, dynamic fracture, and interface and adhesive fracture; and the characterization of material response and failure at the microscale, with an emphasis on experimental mechanics." The Mechanics of Time-Dependent Materials Technical Division has created the SEM/MTDM Wolfgang Knauss Young Investigator Award.

Michael (Mike) Fourney.   Mike is another very important figure on the West Coast of the U.S., and a very active and remarkable contributor to SEM.

Mike's education includes: B.S., Aeronautical Engineering, West Virginia University, 1958; California Institute of Technology, Aeronautics and Mechanics Department, 1959 M.S., Ph.D. 1963.

His academic positions included: University of California at Los Angeles, Department of Engineering and Applied Sciences, Professor 1972–1994; Chairman of the Department of the of Mechanical and Structure, 1979–1983; Chair of the Faculty 1991–1993; Emeritus Professor 1994; prior to 1972 faculty at Washington University in Saint Louis, School of Engineering and Applied Sciences, Mechanical Engineering, and Applied Sciences Department. Mike was also a Visiting Professor to the United States Military Academy, 1989–1990.

Upon his retirement from academia he founded a consulting firm, Fourney Engineering Inc. He has been extensively involved in consulting activities both in governmental and private institutions.

Mike's memberships include: The Society for Experimental Mechanics; American Institute of Physics; Optical Society of America; RILEM (Reunion Internationale des Laboratoires d'Ensais de Materiaux); British Society for Strain Measurement; American Association of University Professors; American Association for the Advancement of Science; U.S. National Committee on Theoretical and Applied Science.

His field of interest is Experimental Mechanics with particular emphasis on applied optics to solid and fluid mechanics problems (photoelasticity, holographic interferometry, speckle techniques, Moiré).

He has published extensively on his field of interest with many papers on dynamic applications of holographic interferometry.

Mike's honors and awards include: SEM Fellow; President 1979; 1990 Murray Lecture; 1996 Tatnall Award; 2001 Honorary Member; 2004, Taylor Award; American Society of Mechanical Engineers, Fellow; Academy of Mechanics, Fellow.

Guruswami (Ravi) Ravichandran   Guruswami (Ravi) Ravichandran's education includes: B.E., University of Madras (Regional Engineering College), 1981; Sc.M. (Solid Mechanics and Structures), Brown University, 1983; Sc.M. (Applied Mathematics), 1984; Ph.D., 1987.

His academic positions include: Assistant Professor, Caltech, 1990–1995; Associate Professor, 1995–1999; Professor, 1999–2005; Goode Professor, 2005–now; Associate Director, 2008–2009; Director, Graduate Aerospace Laboratories, 2009–2015; Otis Booth Leadership Chair, 2015.

His research focuses on the deformation and failure of materials; micro/nano mechanics; wave propagation; composites; active materials; biomaterials; cell mechanics; experimental mechanics; mechanics of lightweight materials; active materials with large strain actuation; and space structures.

His honors and awards include: SEM Fellow; Hetenyi and Lazan awards in 2001; Murray Lecture 2014; President 2016; Doctor honoris causa, Paul Verlaine University, Metz, France; 2013 Eringen medal Society of Engineering Science; Warner T. Koiter Medal from the American Society of Mechanical Engineers (ASME); Chevalier dans l'ordre des Palmes Académiques by the Republic of France, 2015; Member, National Academy of Engineering, 2015; Elected Member academia Europea 2017.

Ravi has an extensive list of publications in top journals: *Journal of the Mechanics and Physics of Solids, Applied Physics Letters, International Journal of Fracture,* and *International Journal of Impact Engineering.* He is co-editor of *SEM Proceedings, Dynamic Failure of Materials and Structures.*

We will now focus on the mid-west (another alma mater of Chuck Taylor's), where the excellence in Experimental Mechanics and the connection with SEM has been continued.

Nancy Sottos.   The Donald B. Willet Professor of Engineering in the Department of Materials Science and Engineering at the University of Illinois Urbana-Champaign.

Nancy's education includes: B.S. and Ph.D. 1986–1991, Mechanical Engineering, University Delaware.

Nancy is currently the Donald B. Willet Professor of Engineering in the Department of Materials Science and Engineering at the University of Illinois Urbana-Champaign. She is also co-chair of the Molecular and Electronic Nanostructures Research Theme at the Beckman Institute for Advanced Science and Technology and a University Scholar.

Nancy's research focuses on the development of autonomic materials systems that have the ability to achieve adaptation and response in an independent and automatic fashion. Recent highlights of Sottos' work include: (i) nanostructured self-healing polymers, (ii) self-sensing, mechanochemically active polymeric materials, (iii) autonomous materials systems with microvascular networks, (iv) fluorescent digital image correlation strain measurement with nanoscale resolution, and (v) metrology for dynamic interfacial adhesion measurement in multilayer thin films.

Nancy is author and co-author of numerous prestigious journal papers and other publications.

Nancy's honors and awards include: SEM-2004 Hetenyi Award; 2011 Lazan Award; 2012 Fellow; President of SEM 2014–2015; Editor-in-Chief *Experimental Mechanics*; 2007 Fellow of the Society of Engineering Science; 2007 Scientific American 50; 2002 University of Delaware Presidential Citation for Outstanding Achievement; 1992 Office of Naval Research Young Investigator Award.

Ioannis Chasiotis.   Professor, Aerospace Engineering, University of Illinois at Urbana-Champaign and Editor in Chief, *Experimental Mechanics*.

Ioannis' education includes: 2002 Ph.D. in Aeronautics (Minor in Materials Science); 1998 Master in Aeronautics both degrees from the California Institute of Technology, Pasadena, California; 1996 Diploma in Chemical Engineering, Aristotle University of Thessaloniki, Greece.

Ioannis' work focuses on the nanoscale mechanical behavior of materials and the associated mechanisms of deformation that make metal, ceramic, and polymeric films and fibers, nanostructured materials, and nanoscale structures stronger and tougher. A key objective of his research is to establish relationships between the material properties and microstructure, and the synthesis and manufacturing processes. He has developed a wide variety of micro- and nanoscale experimental methods to resolve nanoscale temporal and full-field material deformations by using atomic force microscopes, high-resolution optical microscopes, high-speed cameras, or MEMS devices that can test ultra-small fibers.

His professional activities include: Member of Society for Experimental Mechanics; American Society for Mechanical Engineers; Materials Research Society (MRS); Editor in Chief, *Experimental Mechanics* 2016–2020; Guest Editor 2006; International Advisory Board of Experimental Mechanics, 2005–2008; Associate Editor for *Strain*, 2006–2009.

Ioannis has a long list of awards and honors (22 in number) received. SEM, 2013 Durelli and 2012 Hetenyi awards; 2008 Best Research Paper Award; 2015 Fellow of the Society of Mechanical Engineering.

To date, Ioannis has an impressive number of journal publications, 64 in total. Recently, he is the co-editor with Philip Reu of the *Proceedings of the Annual International DIC Society Conference* and *SEM Fall Conference* organized by the Society for Experimental Mechanics and Sandia National Laboratories and held in Philadelphia, PA, 2016.

We continue to the Midwest and its role in the field of Experimental Mechanics and contributions to SEM from Northwestern University and the lineage of M. Hetenyi.

Horacio D. Espinosa. James N. and Nancy J. Farley Professor in Manufacturing and Entrepreneurship, and Director, Institute of Cellular Engineering Technologies (ICET), Northwestern University.

Horacio's education includes: Civil Engineering Magna Cum Laude; six-year professional degree, Northeastern National University, Argentina, 1981; M.Sc. Structural Engineering, Polytechnic of Milan, Italy, 1987, advisor Giulio Maier; M.Sc. Solid Mechanics, Brown University, 1989; M.Sc. Applied Mathematics, Brown University, 1990; Ph.D. Solid Mechanics, Brown University, 1992, *"Micromechanics of the Dynamic Response of Ceramics and Ceramic Composites"* advisors Professors R.J. Clifton and M. Ortiz.

His academic positions include: Purdue University, 1992–1997, Assistant Professor, Aeronautics and Astronautics; 1997–1999 Associate Professor, Northwestern University; 2000–2004 Associate Professor, Mechanical Engineering; 2004–present Professor, Present Director, Theoretical and Applied Mechanics Program; 2009–present James N. and Nancy J. Farley Professor in Manufacturing and Entrepreneurship; 2013–2015 Faculty Director of Nano/Microfabrication core facility (NUFAB); 2015–present Director, Institute for Cellular Engineering Technologies (ICET).

Horacio achieved worldwide recognition for contributions to the experimentation and modeling of: dynamic failure of materials; lab-scale fluid-structure interaction experiments; unraveling the mechanics of biomaterials; discovering size effect dependent mechanisms and mechanical properties of freestanding thin films and nanomaterials, e.g., carbon nanotubes and nanowires, the pioneering of robust carbon-based nanoelectromechanical systems (NEMS); and the creation of microfluidic platforms for sub-hundred nanometer patterning and single-cell manipulation and analysis. The achievements of Horacio and his research group results in two outstanding conclusions: the depth of the theoretical knowledge displayed and the extraordinary skills in implementing experiments to support and enhance theory and create instrumentation for practical applications.

A prolific writer, Horacio has an impressive record of publications in many different types of scientific and technological literature totaling 170 papers where he is an author or co-author. These publications reflect his outstanding achievements in the different topics of his research field.

Horacio has 10 U.S. patents and disclosures.

His honors and awards include: SEM President, 2010; Editor-in-Chief *Experimental Mechanics*; Hetenyi Award, 2005; 2008 Lazan Award; 2008 Fellow; 2013 Sia Nemat-Nasser Award; 2016 Murray Lecture. American Academy of Mechanics; 2001 Fellow; 2012 President Society of Engineering Science; National Academies Panel on Materials Science and Engineering to advise the Army Research Lab, 2013-present; Fellow, American Academy for the Advancement of Science (AAAS), 2013; Thurston Lecture Award, American Society of Mechanical Engineers, 2015.

We turn now our attention to other parts of the Midwest that have been centers of excellence in fields connected to Experimental Mechanics and that have provided many contributions to SEM.

5.1 MEMBERS FROM UNIVERSITIES IN MICHIGAN: MICHIGAN STATE UNIVERSITY, OAKLAND UNIVERSITY, AND MICHIGAN TECHNOLOGICAL UNIVERSITY

Gary L. Cloud.   Distinguished Professor of Mechanical Engineering, Michigan State University.

Gary's academic position include: 2007 Distinguished Professor of Mechanical Engineering, serving in different positions for 57 years; Director Emeritus M.S.U Composite Vehicle Research Center.

Gary's education includes: 1959 B.S. Michigan Technological University; 1961 M.S. Michigan Technological University; 1966 Ph.D. Michigan State University.

Gary's research interests involve bringing together optical and electronic techniques with analytical mechanics to solve interesting problems in geomechanics, biomechanics, composites, fracture mechanics, fastening, and nondestructive evaluation.

He has authored and coauthored about 200 papers and has published in 26 different archival journals: *Optical Methods of Engineering Analysis Cambridge*, University Press 1995, second printing (1998); chapters in the *Handbook of Experimental Mechanics* and the *Marks Mechanical Engineering Handbook*. In 1992, papers on "Back to Basics - Optical Methods" for the journal *Experimental Techniques;* the series spanned 10 years and ran for 50 installments.

Gary's professional activities include: President of SEM 1993–1994; SEM, Technical Editor *Experimental Mechanics* from 1985–1988; Associate Editor for eight years; Chair of the Editorial Council, Board Member, and in many other posts. He has been a member of SEM for over 50 years. In 2008, SEM created the "Gary L. Cloud Scholarship Endowment" to assist students who intend to make a career in experimental mechanics. Dr. Cloud served on the International Editorial Board of the journal *Measurement Science and Technology* for several years. He holds membership in several other technical and honorary organizations, including the Society of Photo-optical Instruments, the Society of Automotive Engineers, the British Society for Strain Measurement, and the Institute of Physics.

Gary holds three patents and has another patent pending.

His awards and honors include: SEM Fellow Award; 1990 Harting Award; 2000 Frocht Award; 2006 Tatnall Award; 2012 Murray Lecture; 1994 "Distinguished Faculty Award" from Michigan State University; 1995 "Withrow Senior Distinguished Scholar Award" from the M.S.U College of Engineering; He is a two-time recipient of the "John and Dortha Withrow Excellence in Teaching Award." In 1992, he received the "Outstanding Engineer Award" from the Michigan Society of Professional Engineers for his work in engineering education. Since 1995 he has been a Faculty Advisor of the M.S.U Formula SAE Racing Team, and he has worked to bring significant improvements in professionalism and performance to the team.

We cannot fail to mention Gary's thesis advisor, Jerzy-Tadeusz Pindera, Professor of the University of Waterloo and also an active member of SEM. Jerzy is well known for his developments in a subject that he called Integrated Photoelasticity and the method of isodynes. A remarkable individual, he was a survivor of the Holocaust. In 1940, he was sent for extermination to the concentration camp, Konzentrationslager Sachsenhausen, by the Gestapo. By a miracle he survived and regain freedom after a dead march of prisoners when American and Soviet troops advanced inside Germany. He narrated to me all his experiences and described in vivid terms the extreme cruelty of the SS German officers.

Joseph Der Hovanesian. Joseph was Chair of the Department of Mechanical Engineering at Oakland University.

Joseph's academic positions included: Assistant Professor and postdoctoral Pennsylvania State University, 1958–1960; Associate Professor Wayne State University, Detroit, 1960–1970; Professor of Engineering Oakland University, Rochester, Michigan, from 1970; Department Head Mechanical Engineering, from 1973. He was the director of a series of courses supported by National Science Foundations on applied Optics and Holography.

His education included: 1950 B.S. in Science, 1953 M.S. of Arts; 1958 Ph.D., Michigan State University.

His fields of research were optical techniques, photoelasticity, holography, speckle techniques, projection Moiré applied to experimental mechanics particularly with applications to problems of the automotive industry including analysis of deterministic approach (factor of safety), and statistical methods.

He published a large number of papers on the topics of his interest, co-authoring many of them with Mike Hung a close friend and collaborator of Joe's.

Joseph's honors and awards included: 1988, SEM Fellow; 1997 Frocht Award.

We move now to other members of the SESA/SEM and Universities and Centers of Excellence where they worked.

John Ligon. John is currently a Professor Emeritus of Mechanical Engineering and Engineering Mechanics, Michigan Tech.

John was born in Texas. He earned his B.S. in 1964 and M.S. in 1965, both in Mechanical Engineering from Texas Tech University. He earned his Ph.D. in 1971 in Engineering Mechanics from Iowa State University.

Having been in Army ROTC and a commissioned officer he had a deferment while in graduate school and served on active duty from August–December 1970 at Aberdeen Proving Ground, MD. He worked for Bechtel Corporation in San Francisco as a Senior Engineer from 1971 until he joined Michigan Tech in 1972 as an Assistant Professor. His professorial career was at Michigan Tech where he was promoted to Full Professor in 1981. John retired from Michigan Tech in September 2006 after 34 years of dedicated service.

John's awards include the SESA 1976 R.E. Peterson Award for the most outstanding applications paper in the *Journal of SESA* over a two-year period, the SEM 1990 M.M. Frocht Award for outstanding achievement in education of experimental mechanics, and the SEM 1993 Tatnall Award for long and distinguished service to the Society. He was President of the Society for Experimental Mechanics (formerly SESA) 1983–1984. He was elected to the Texas Tech University Academy of Mechanical Engineering in 1995. While John was President of SESA he led a campaign to change the name of the Society for Experimental Stress Analysis to the Society for Experimental Mechanics. He was elected fellow of the SEM in 1997.

John was an active researcher with external funding from a wide variety of government and industrial sponsors. He was the co-developer of the field of Phytomechanics, which is concerned with applying experimental methods to the study of plants. He has graduated 31 M.S. and Ph.D. students and published over 40 technical papers.

John has also had a long and distinguished career of Engineering Mechanics education on many fronts. He is a respected teacher by his students. In 1991, he organized and chaired the SEM's first International Student Paper Competition and continued as chair for many years.

5.2   OTHER MEMBERS AND UNIVERSITIES OF NOTE ACROSS THE UNITED STATES

C.W. "Bill" Smith (2012).   Bill was an unforgettable colleague, a living example of a southern gentleman, always with his characteristic smile and his amiable demeanor.

His education included: B.S., M.S., Ph.D., from Virginia Tech.

His academic positions included: Instructor in 1948; Assistant Professor in 1950 and Full Professor at the Department of Mechanic; 1992 Distinguished Professor Emeritus of Engineering Science and Mechanics and a member of the Academy of Engineering Excellence at Virginia Tech.

When George Irwin gave a seminar at Virginia Tech, Bill became interested in the field of fracture mechanics and devoted all his research effort utilizing Photoelasticity as a tool in many aspects of his research. His work in fracture mechanics was outstanding and earned him a worldwide reputation.

Bill authored or co-authored more than 150 papers, contributed to many book chapters. He was the editor of the *Journal of Fracture Mechanics* and *Theoretical and Applied Fracture Mechanics*.

His awards and honors included: being an active member of SEM; in 1977, he was elected Fellow; in 1983 he received the Frocht Award; in 1993 he was awarded the Murray Lecture; in 1995 the Lazan Award; and in 2002 he was elected as Honorary member. In 1986, he received NASA's Langley Research Center Scientific Achievement Award. Other honors followed including election to Fellow of the American Academy of Mechanics in 1991 and of the American Society of Mechanical Engineers in 1996.

William (Bill) Sharpe.   Professor Emeritus, The Johns Hopkins University.

Bill's education includes: North Carolina University, B.S. 1960; M.S. Ph.D. 1961, The Johns Hopkins University.

His academic positions include: The Johns Hopkins University. Professor of Mechanical engineering; 1983 Chair of the Department of Mechanical Engineering; he is currently Professor Emeritus.

His research work is in the field of experimental solid mechanics, microelectromechanical systems (MEMS); and microsample testing.

He has published extensively in his areas of interest particularly in the determination of properties of micro-sized specimens filling the gap between macro to micro properties and fracture. He is editor, Springer *Handbook of Experimental Solid Mechanics*.

His honors and awards include: SEM President 1984; Fellow – ASME Award; Frocht Award; Murray Lecture; Tatnall Award; Lazan Award; 2012 Honorary Member SEM; Best Paper - *Journal of Engineering Materials and Technology Materials Division of ASME* - vol. 105(1983); Fellow – ASME; Nadai Award—ASME, 2007; Roe Award from ASEE.

Mark Tuttle.   Mechanical Engineering, University of Washington, Seattle, Washington.

Mark's education includes: B.S., Michigan Tech. Mechanical Engineering 1975; M.S. Engineering 1984; Ph.D. from Virginia Tech.

His academic positions include: Mechanical Engineering, University of Washington, Seattle, Washington State; 1985 Assistant Professor; 1990 Associate Professor; 1995 Full Professor and Chair of the Department 2004–2010; Adjunct Professor of Industrial Engineering. Currently, Mark is the Director of the Center for Advanced Materials in Transport Aircraft Structures (AMTAS).

His research work is in the area of applied solid mechanics, composite materials and structures, adhesion mechanics, and viscoelasticity. Mark's studies have been devoted to predicting the mechanical response of discontinuous- and continuous-fiber composites, developing new composite repair technologies, the buckling response of composite laminates, optimal design of composite structures, and prediction of the long-term durability of composites.

Mark has published a large number of papers in the area of composites and processes of joining and repairing composites.

His honors and awards include: SEM President (1995–1996); 1982 Harting Award; 2009 Tatnall Award; 1984 Sigma Xi Ph.D. Research Award.

Arkady Voloshin.   Lehigh University Bethlem, PA.

Arkady's education includes: Dipl., Solid State Physics, Leningrad Polytechnic (USSR); Ph.D. in Experimental Mechanics, Tel-Aviv University, Israel.

His academic positions started in 1984 as a Professor in the Department of Mechanical Engineering and Mechanics at Lehigh University Bethlehem, PA.

Arkady was a Visiting Professor at École Nationale Supérieure des Mines de Saint-Étienne; Center for Biomedical and Healthcare Engineering; University of Ljubljana; Faculty of Natural Sciences and Engineering Slovenia, Ljubljana; Pontificia Universidade Catolica de Minas Gerais in Brazil; and Aoyama Gakuin University in Japan.

His research work includes: image processing, pattern recognition, computer vision, digital image processing, fracture mechanics, cell and tissue engineering, and biomechanics.

He has published a large amount of papers (143) in the subjects of interest and has 2,722. citations. He is co-author with Igor Emri, *Statics, Learning from Engineering Examples*, Springer, 2016.

His honors and awards include: recipient of a M. Hetenyi and Brewer Awards from the SEM; elected as an Academic Advisor by the Presidents of the International Academy of Science in Russia for his "significant scientific achievements in the solution of the engineering problems of mechanics."

Sanishiro Yoshida.   Southeastern Louisiana University, Department of Chemistry and Physics, Professor of Physics/Integrated Science.

Sanishiro's education includes: Keio University of Science and Technology, Department of Electronics and Electrical Engineering; 1980 B.S, 1983 M.S, 1986 Ph.D.

His academic positions include: 2001 Professor Southeastern Louisiana University, Department of Chemistry and Physics, Professor of Physics/Integrated Science; 1981–1984 Instructor Keio University; California Institute of Technology, 2001–2011 Visiting Professor working in the development of the interferometer of the LIGO project and teaching Physics courses; University of Florida, 1997–2000 Physics Department Adjunct Professor; R & D Center for Applied Physics, Indonesian Institute of Sciences, Senior research Scientist, December 1993–April 1997; Hutech Research Laboratory, Russia Director of Technical Division, October 1990–November 1993; Institute for Laboratory Astrophysics, Univ. of Colorado, Boulder, Visiting Scientist, 1981–1982.

His research interest has covered a wide range in the field of electrical engineering related to laser technology. From his laser technology work, he became interested in speckle interferometry as a tool to pursue the experimental aspects of a theory that he started to work on during his appointment in Russia called gauge theory. This work was fundamental in establishing a connection between Maxell's equations and solid-state matter organization. He worked also on interferometry problems of the LIGO project. A very important portion of his research work has been focused on the plastic deformation of materials based on meso-mechanics and the experimental verification of theoretical predictions using speckle interferometry. This work has established an important aspect that the classical Theory of Plasticity has ignored; the dynamic aspect of plasticity and its connection with wave-propagation even at very slow rates of loading.

He has published a large number of papers presenting successive stages of the development of what he calls gage-theory of solids.

His awards and honors include: Southeastern Louisiana University President's Award for Excellence in Research 2008; recognized as one of the 50 innovators in the New Orleans area 2008; Southeastern Faculty Senate Research Award 2016.

Ryszard J. (Rich) Pryputniewicz.   Worcester Polytechnic University (WPI).

In describing the WPI contribution to SEM, I must include a friend and colleague that due severe health problems is no longer professionally active but that has made many important contributions to the profession and to SEM. Many years ago, my good friend Karl Stetson contacted me about helping a very promising young holographer that he had met in Europe and wanted to get away from the Communist regime prevailing at that time in Poland. Karl's efforts resulted in Ryszard coming to the U.S. He was at WPI as the K.G. Merriam Distinguished Professor of Mechanical Engineering as well as Professor of Electrical and Computer Engineering, and, since 1978, founding Director of the Center for Holographic Studies and Laser micromechatronics (CHSLT). He was also founding Director of the Nano-Engineering, Science, and Technology (NEST) Program at the Mechanical Engineering Department of WPI. Prior to joining WPI in 1978, Rich was a member of the faculty and Director of the Laser Research Laboratory at the School of Engineering and the Schools of Medicine and Dentistry of the University of Connecticut (six years) and a member of the technical staff in the aerospace industry (four years). A holographer by training, his research and teaching interests concentrated on theoretical and applied aspects of MEMS, smart sensors and structures and nanotechnology, lasers, noninvasive metrology with nanometer accuracy on sub-micron scale, and nondestructive testing (NDT) based on photonic methods with emphasis on heat transfer, thermal management, and design optimization. In this work, he emphasized the unification of analytical, computational, and experimental solutions (ACES) methodology, which he pioneered, especially when it can be merged to provide results where none would be obtained otherwise, to ease the solution procedure, or to attain improvements in the results. He became a Member of the European Academy of Sciences and Arts (EASA), Fellow of SPIE, Fellow of SEM, Fellow of ASME, Senior Member of IEEE, Chairman of the Development Committee of the MEMS Division of ASME, President of SEM, and chairman of the Education Committee of the IEEE Nanotechnology Council. He has written over 400 papers. A very active member of SPIE and organizer of many symposia in the 1990s, upon my invitation he joined SEM and put a bulk of his efforts into our Society. He received the Frocht Award and the Murray Medal in 2008 and became President in 2010.

Cosme Furlong.   Worcester Polytechnic Institute, Worcester, Massachusetts.

Cosme's education includes: B.S. 1989 University of Americas Mexico; Worcester Polytechnic Institute 1992 M.S; 1999 Ph.D. Mechanical Engineering.

His academic positions include: Worcester Polytechnic Institute—Assistant Professor, Associate Professor, currently Full Professor; head of the Center for Holographic Studies and micro-mechatronics. Associated appointments with Massachusetts and Eye and Ear Infirmary and Harvard Medical School.

Cosme's research interests include the research, development, and implementation of optical metrology and nondestructive testing techniques, fiber optic sensors, and computer vision and image processing algorithms for applications pertaining to the combined use of analytical, computational, and experimental methodologies for the study and optimization of mechanical, electro-mechanical, optical, and MEMS/MOEMS components and packages.

He has authored or co-authored papers in his areas of interest utilizing holographic interferometry as a main tool. His current work has introduced advances in holographic interferometry to solve problems in audiology and design of actual equipment for actual in vivo observations.

His honors and awards include: 1988 Mechanical Engineering Outstanding Student Award; 1999 Sigma Xi Ph.D. Research Award, Worcester Chapter; SEM 2007 President MENS and Nanotechnology Division; 2010 Durelli Award; 2018 President Optical division.

Felix Zandman (1928–2011).   Polish-born entrepreneur and founder of Vishay Intertechnology, one of the world's largest manufacturers of electronic components in the world. From 1946–1949, Felix studied in France at the University of Nancy in the Departments of Physics and Engineering. Simultaneously, he was enrolled in the Grande École of Engineering Ensem (École nationale supérieure d'électricité et de mécanique). He received a Ph.D. in Physics at the Sorbonne on a subject that became a central point in his life…photoelastic coatings.

I want to stress an important event that took place in his life that he narrated to me in very vivid terms. Felix described to me the dramatic events that took place in a barn when he and some relatives were almost caught by the Nazi's Gestapo. He told me that he could hear the voices of the German officers. He did not die in the Holocaust due to the efforts of a Catholic family that shielded him for many months. The advancing Soviet Army liberated the area where he was hidden. In 1946, he was able to immigrate to France. Eventually, he emigrated to the U.S. and joined Frank's Tatnall's company in Philadelphia as Director of Basic Research. In 1962, he founded his own company Vishay Intertechnology. Vishay has become a Fortune 1000 company with many subsidiaries and over 22,000 employees worldwide. Vishay Intertechnology (NYSE: VSH) is a publicly traded company with a market capitalization of over a billion dollars. Felix's life resulted in quite a remarkable achievement, bringing a capability to blend scientific and technological knowledge with business skills. Photoelastic coatings became the successors of stress coats and Zandman, with his staff and collaborators, developed specialized photoelastic equipment and material that has been extensible utilized as a tool to obtain field information at surfaces. From photoelastic coatings he moved to the strain gage field where his company became one of the main world fabricators of foil gages and the required electronic equipment to perform measurements. He was the author or co-author of many papers about photoelastic coatings. He was the co-author of a book titled *Photoelastic Coatings*, with Salomon Redner and James W. Dally, that was first published in 1977 and reprinted in 2006. He co-authored a book, *Resistor Theory and Technology*, with Paul-Rene Simon and Joseph Szwarc. In May 1989, the SEM Executive Board formally created the Felix Zandman Award. The award honors Felix for his extensive work with photoelastic coatings and his support of SEM. In 1996 he was made an Honorary member of the Society.

Archie A.T. Andonian.   Senior R&D Scientist for Calnetix Technologies, Retired from Goodyear Tire Company Research Labs.

Archie is and has been a very active member of SESA/SEM for 35 years and has been a member of the more important positions in the governance of the Society including SEM President in 2009.

Archie received his Ph.D. in Engineering Science and Mechanics from Virginia Polytechnic Institute and S.U. in 1978.

He has published in *Engineering Fracture Mechanics*, *Applied Optics*, *Experimental Mechanics*, *Experimental Techniques*, *Journal of Biomechanics*, *ACI Journal*, and *Journal of Material Science*, and has more than 200 research papers. Recently he received the SEM Fellow Award.

Now we shift our attention from outstanding members of other communities that also have interacted and contributed to SEM in the fields of theoretical and applied mechanics.

Ares Rosakis.   A native of Greece, in 1975 he studied Engineering Sciences at the University College Oxford.

He received his B.A. and M.A. in 1978 and 1986, respectively. He earned his ScM. (1980) and Ph.D. (1982) in Engineering, Solid Mechanics, and Structures from Brown University. In 1982, he joined the California Institute of Technology (Caltech) as an Assistant Professor. In 1988, Ares became Associate Professor and Full Professor 1993, respectively. In 2004, he was named the Theodore von Kármán Professor of Aeronautics and Professor of Mechanical Engineering. In 2013, he was honored as the inaugural recipient of the Otis Booth Leadership Chair of the Division of Engineering and Applied Science, equivalent to the position of Dean in other universities. It takes pages to enumerate all the awards and honors he has received. The awards that he received from SEM: Fellow, the Hetenyi, the Lazan, and Theocaris awards. Ares has contributed to the field of quasi-static and dynamic failure of metals, composites, and interfaces using high-speed visible and infrared diagnostics and laser interferometry. Ares combined engineering fracture mechanics and geophysics to gain a better understanding of the destructive potential of large earthquakes, a field that he has dedicated a great deal of efforts. An outstanding contribution to geophysics was his experimental discovery of "intersonic" or "supershear" ruptures or dynamic delamination cracks. These ruptures are capable of propagating at speeds that are faster than the shear wave speeds of the surrounding material, and can spread along fault planes in the earth's crust to produce supershear earthquakes. These ruptures also grow along weak interfaces in a variety of composite materials commonly used in engineering practice.

He holds 13 U.S. patents on thin-film stress measurement and *in situ* wafer-level metrology as well as on high-speed infrared thermography. He authored of more than 260 papers on the dynamic deformation and catastrophic failure of metals, composites, interfaces, and on laboratory seismology. He is a member of the National Academy of Engineering and a fellow of the American Academy of Arts and Sciences.

Sia Nemat-Nasser.   Sia is a native of Iran and in 1958 he emigrated to the U.S. He earned his B.S. from Sacramento State University and his M.S. and Ph.D. from UC Berkley, in 1961 and 1964, respectively. He joined Sacramento State University as an Assistant Professor 1961–1962. He moved to Northwestern University as Post-Graduate, where he became Assistant Professor, Associate Professor, and Full Professor in 1966. He left Northwestern in 1985 to join the University of California at San Diego. He is currently the Director of the Center of Excellence for Advanced Materials. Current research interests include: micromechanical and constitutive modeling of nonlinear response and failure modes; analytic and computational mechanics; and static and dynamic experimental characterization of materials, especially advanced composites, ceramics, and ceramic composites; advanced metallic and polymeric composites particularly polyelectrolytes and ionic polymer-metal composites, and high-strength alloys, as well as rocks and geomaterials. An outstanding contributor to the field of solid mechanics both theoretically and experimentally, he is a prolific writer in the fields of interest. He is the author of several books, 1993, *Micromechanics*, 1997 *Dynamic Behavior of Brittle Materials*, 2004 *Plasticity*, and editor *Mechanics Today: Pergamon Mechanics Today Series*, 2013. He has received many honors and awards: National Academy of Engineering, Member 2001; ASME: Stephen P. Timoshenko Medal 2008; Mater. Div. established the Sia Nemat-Nasser Early Career Medal in 2008 (focused on underrepresented minorities & women in engineering); Robert Henry Thurston Lecture Award 2006; Honorary Member 2005; Aerospace Division, Adaptive Structures & Materials Systems Best Paper of the Year Award 2003; Nadai Medal 2002; Life Fellow 2001; Chair, Mater. Div. 1997–1998; Fellow 1979; ASCE: Theodore von Karman Medal 2008; Sacramento State University: Distinguished Alumni Award 2008; SEM: B.J. Lazan Award 2007; W.M. Murray Medal 2009. SES: William Prager Medal 2002; Founding Fellow 1988; Pres. 1979–1980. AAM: Pres. 1996–1997, Secretary 1989–1994, Fellow 1970. World Innovation Foundation: Honorary Member 2004; Tech. Inst.: Willard F. Rockwell Medal 2003; UCSD: Faculty Research Lecturer Award 2005; Sch. Eng./MAE Teacher of the Year Award, 2000–2001, 1996–1997, 1994–1995; John Dove Isaacs Chair in Natural Philosophy, 1995–2000; Alburz Educational Foundation Prize 1975; SEM created the Sia Nemat-Nasser Award.

5.3   FROM THE FIELD OF THEORETICAL AND APPLIED OPTICS

Karl A. Stetson.   Experiments performed from October–December 1964 by Karl and R.L. Powel lead Karl to the explanation of the formation of holographic interferometry patterns, when he was a Master's Degree student working at the Radar Laboratory of the University of Michigan's Institute of Science. This discovery resulted in a paper published in the *Journal of the Optical Society of America* in 1965. This event was the beginning of Karl's life dedication to this discovery and the introduction and development of one of the more important tools of Experimental Mechanics. He received his Doctoral Degree at the Royal Institute of Technology in Stockholm and worked most of his life at the United Aircraft Laboratory, East Hartford Connecticut. Upon retirement he started his own company, Karl Stetson Associates, that manufactures digital holography systems to measure vibratory and static deformations of engineering structures as well as shape and strain measurement. He is the author of many seminal papers in different topics pertaining to the theory of formation of holographic fringes, their interpretation, and applications. Many years ago, Karl and I met and shared a common interest in the field of optics and its application to the field of mechanics. I always admired him as an individual that had accumulated so many achievements in science and technology and had an interest in music. I have always admired his strong knowledge in the mathematics and optics fields and at the same time his capability of using these tools to achieve many experimental outstanding successes such as obtaining, in the field, holographic fringe patterns in a working jet engine. He received the following SEM award: 1999 Murray Lecture.

CHAPTER 6

SESA/SEM Active Members Abroad

The society has its International Congress every four years during which a large group of foreign researchers participate. Besides this general group, there are foreign groups and individuals that regularly attend SEM meetings and participate actively in the Society's organization and other tasks. The following people have made very important contributions to the Society.

Janice Barton. Professor of Experimental Mechanics, Engineering, and the Environment at the University of Southampton. Professor Barton is very well known for her outstanding research in thermography/thermoelastic stress analysis areas. She was awarded SEM Fellow status in 2016.

Igor Emri. From 1996, Igor has been the Professor and Head of the Department of the University Ljubljana, Faculty of Mechanical Engineering, Slovenia. He is known worldwide for his contributions to: mechanics of time-dependent materials and both linear and nonlinear viscoelasticity; effect of temperature, pressure, and humidity on mechanical properties of polymers and composites; fatigue and fracture mechanics of polymers and composites; general and polymer rheology; polymer processing, composite manufacturing, and processing; new experimental methods; dynamic and static analysis of materials and structures; adaptronics; multifunctional and intelligent materials; biopolymers; biomechanics; medical engineering; mechanics of dissipative systems; and environmental engineering. Igor is another of the foreign active participants of the Society activities. He has an education connection with the U.S. having earned his Ph.D. at the California Institute of Technology from the Department of Aeronautics and Applied Mechanics, his thesis advisor being Wolfgang Knauss. The Mechanics of Time-Dependent Materials Technical Division (MTDM) of the Society for Experimental Mechanics was established in 1993 under the initiative of Knauss and Emri, as a response to the needs of the aviation and automotive industry for a better understanding of time-dependent structural behavior of polymers and their composites at extreme thermo-mechanical loading. Dr. Knauss and Dr. Emri were the first chair and co-chair of the newly established Technical Division. The news was announced in the January/February 1993 issue of the *Experimental Techniques*. Since then, MTDM is one of the most active SEM Technical Divisions and Igor one of the more active participants and organizers of symposia and other activities of this Technical Division.

Igor was Vice President 1998–1999, President 1999, member of the Executive Board 1955–1997, 1996 Section of the Year Award, and 2009 Fellow.

Jose Freire.   Professor of the Pontifical Catholic University of Rio de Janeiro School of Engineering and Chairman of the Photomechanics Laboratory. Jose is part of the group of consistent members of SEM that always shows up at meeting and that serve SEM in different capacities, i.e., as a member of the Executive Board of the Society, Chairman of Editorial Council of the Society for Experimental Mechanics (2005–2007), Secretary, Vice Chairman and Chairman of Optical Methods Division of the Society for Experimental Mechanics (1992–2001), and President of the Society 2004–2005. He is also connected to the U.S. Education System having received his Ph.D. from Iowa University in 1979. He received the Zandman Award 1998, Fellow Award, SEM, 2011, and Tatnall Award, SEM, 2013.

Emmanuel Gdoutos.   Professor Director of the Laboratory of Applied Mechanics, Director of the Section of Design and Construction of Structures Department of Civil Engineering, Democritus University of Thrace, Greece. It takes many pages to outline all the contributions Emmanuel has given to the fields of solids mechanics, fracture mechanics, and experimental mechanics in Europe and the U.S., and all the recognition that he has received for these contributions. Emmanuel is one of the individuals that has consistently attended SEA/SEM meetings and participated actively in the Society's governance and service.

President, 2013–2014, Vice President, 2011–2012, Member of the Honors Committee, 2005–2008, Member of the Fellows Committee, 2006–2008, Chairman of the Fellows Committee, 2007–2008, Member at Large of the Executive Board, 2006–2008, and Member of the Editorial Board of *Experimental Mechanics*. Awards of the Society: Fellow Award, 2009 Theocaris Award, Lazan, 2010 Tatnall Award, and 2011 Zandman Award.

Michael Grediac.   Professor Grediac is from the Department of Mechanical Engineering, Institute Pascal, Clermont-Ferrand University. His research interests include the use of full-field measurement techniques in experimental solid mechanics. He is a very active member of SEM and has received from SEM the following awards: 2010 Theocaris, 2011 Hetenyi, and 2015 Lazan. He is currently an associate editor of *Experimental Mechanics*.

Francois Hild.   Professor Hild is from LMT, ENS Paris-Saclay/CNRS/University, Paris-Saclay, France. A steady attendant of SEM, Francois is world renowned for his seminal work in digital image correlation, and its methodology and applications. He received the SEM Lazan Award in 2014.

Luciano Lamberti.   Associate Professor of the Politecnico di Bari Diapartimento di Meccanica, Matematica E Management. He is a close collaborator of mine and we have published many papers together in the topics of optical techniques applied to solid mechanics, papers that have been presented in the SEM annual meetings since the earlier 2000s and have resulted in many journal papers. He is a world-renowned authority in structural optimization where he is editor and co-editor of journals in this discipline. Besides his work in experimental mechanics and optimization techniques, he has done seminal work in material's constitutive equations mathematically representing the results of experimental observations done with optical techniques, atomic force microscope, and micro-indentation procedures. He has applied these developments to characterize the mechanical properties of biomaterials. He has published more than 200 papers and has more than 1200 citations of his work. Since the earlier 2000s he has attended the Society annual meetings, and has been a very active member of the Optic Division becoming Secretary, Vice President, and President and organizing many of the Optical Division meetings and editing the corresponding Society Proceedings. He is also a Technical Editor of *Experimental Techniques.*

Yoshiharu Morimoto. Professor Emeritus at the Wakayama University. He is very well known for his numerous contributions to the field of Moiré technologies. He was a member of the Executive Board in 2008 and became a SEM Fellow in 2010.

Eddie O'Brien. A graduate of the University of Sheffield, Eddie retired from Airbus and is a leader in the Aircraft design field. Currently an Independent Consultant of aircraft structures. President 2003–2004, and a Fellow 2012. He received the Zandman, Brewer, and Tatnall Awards in 1992, 2008, and 2011, respectively.

Carmine Pappalettere. Professor Pappalettere is from the Department of Mechanical Engineering, Department of Mechanics, Mathematics, and Management, Politecnico di Bari. For many years, Professor Pappalettere was an active member of SEM and an author or co-author of more than 400 of papers in different areas of Experimental Mechanics with 1,755 citations. Members of his research group include Katia Casavola, Claudia Barile, and G. Pappalettera, all of whom are also frequent participants at SEM conferences. He received the SEM Fellow Award in 2011.

Fabrice Pierron. Professor Pierron is a Professor of Solid Mechanics within the Engineering and the Environment Department at the University of Southampton. His research interest lies in the general areas of the mechanics of deformable solids and failure of engineering systems. Understanding, modeling, and testing this behavior is therefore of primary importance to many engineering areas such as transportation, health (biomechanics), infrastructure, and microelectronics. His research deals with a very wide range of materials–from engineering metals (including welds) to polymers, composites, wood, foams, concrete, tissues, etc. Fabrice became an SEM Fellow in 2013. Fabrice is currently Editor-in-Chief of *Strain.*

Masa Takashi. The late Masa, as we used to call him, was an Engineering Professor at the University of Tokyo. He was an outstanding experimental mechanician with contributions in many different areas, particularly in optical methods applied to different topics of mechanical of materials. He presented many of his contributions at the SEM meetings. He was SEM President, 2007–2008.

Wei-Chung Wang.   Professor National Tsing Hua University Taiwan, Republic of China. Wei-Chung is one of Asia's prominent figures in the field of experimental mechanics with emphasis in optical techniques, fracture mechanics, and non-destructive evaluation. He also has ties to the U.S. university system having received his Ph.D. in Engineering Mechanics from Iowa State University in 1985. He and his co-workers are active participants in the Society's activities. He was President of SEM from 2009–2010, served on the Executive Board, and was Associate Technical Editor of *Experimental Mechanics* (2006–2009). He is a Fellow of the Society and received the Zandman Award and the Tatnall Award in 2015.

CHAPTER 7

The Role of the U.S. National Laboratories

Our attention now focuses on the contributions from our U.S. National Laboratories.

7.1 SANDIA NATIONAL LABORATORY, ALBUQUERQUE, NEW MEXICO

Sandia National Laboratories is a multi-mission laboratory currently operated by National Technology and Engineering Solutions of Sandia LLC, a wholly owned subsidiary of Honeywell International Inc., for the U.S. Department of Energy's National Nuclear Security Administration. Members of this institution have been and are active members of SEM.

Michael J. Forrestal. Retired as Distinguished Member of Technical Staff from Sandia National Laboratories. A Ph.D. from the Mechanics Department of the Illinois Institute of Technology has been an active member of SEM's Division of Dynamic Behavior of Materials and has published many SEM papers on this subject. He received the Peterson award in 2012.

Jonathan (Jon) Rogers. Jon is currently Treasurer of SEM…an extremely important officer and member of the Society. Jon received his B.S., M.S., and Ph.D. degrees in Engineering Mechanics from Iowa State University in 1980, 1984, and 1986, respectively. Jon joined Sandia in the fall of 1986 in the Vibration Testing Division. In the test organization, he worked as the test engineer for vibration and shock testing on a number of systems. Jon was the project leader for the VIBRAFUGE development project which placed a 4000-lb. force rated shaker on the 29-ft underground centrifuge, and for the Acoustic Test Facility development project. This resulted in the construction of the 16,000-cu. ft high-level chamber with combined acoustic and vibration test capabilities.

Jon moved to Systems Studies in the Fall of 1992. He has worked on a variety of studies including Advanced Manufacturing, the Impact of Technology on the Economy, and many studies involving the weapons program and Underground Facilities. Jon was made a Distinguished Member of Technical Staff in the Fall of 2002 and was promoted to manager in the Fall of 2003. He is currently a Senior Scientist in the Systems Analysis and Decision Support Group.

Jon has been an active member of the SEM since 1981. He has served many roles for the Society, including: Member of the Executive Board, Chairman of the Technical Program for the Annual Meeting (4 times), President of the Society, Associate Technical Editor of Experimental Techniques, Chairman of the Editorial Council, and Treasurer of the Society.

Helena Jin.   Helena Jin received her bachelor's degree in mechanical engineering from the University of Science and Technology of China (USTC) and her Ph.D. in mechanical engineering from the University of Maryland College Park. Helena is a technical staff member at Sandia National Laboratories California. Her research interests are experimental mechanics and mechanics of materials, with special focus on developing experimental techniques for mechanical characterization at multi-scale and multi-rate, coupling with various optical methods. Helena joined the SEM community when she was a graduate student and she has been actively involved in the society ever since then by organizing sessions and tracks, leading the optical methods technical division (Secretary, Vice Chair, and Chair 2010–2016) and research committee (Secretary, Vice Chair 2015–). She was elected as Member at Large of the SEM Executive Board from 2016–2018. Helena loves this cozy, friendly, and family-like, but well-organized, SEM community. She always encourages graduate students to get involved in the Society as early as they can.

Phillip L. Reu.   Phil is a Principal Member of the Technical Staff at Sandia National Laboratories. He obtained a Master´s degree in biomedical engineering from Rensselaer Polytechnic Institute and masters and Ph.D. degrees from the University of Wisconsin at Madison in Mechanical Engineering. Since 2003 he has been working in the field of optical measurement techniques, specializing in the areas of Digital Image Correlation (DIC) and coherent laser measurements. Current research efforts in DIC are focused on uncertainty quantification. Phillip is the author of the "Art and Application of DIC" in the journal of *Experimental Techniques*, international instructor in DIC techniques for "Metrology Beyond Colors," and chair of the DIC Challenge. Other areas of active research include pulsed holography and electron Doppler velocimetry for nano-dynamics. His current job is full-scale testing at Sandia, where he is developing new techniques for large-scale and high-rate full-field measurements for application to explosively driven events. Phil is a very active member of SEM and an initiator of the DIC Challenge. The purpose of the DIC challenge is to supply the image correlation community with a set of images for software testing and verification. He was awarded the Brewer award as an outstanding experimentalist from SEM in 2016.Current research efforts in DIC are focused on uncertainty quantification. Phillip is the Vice President of the International Digital Image Correlation Society and co-editor of the *SEM/International Digital Imaging Correlation Society: Proceedings of the First Annual Conference*, Springer 2016.

7.2   LOS ALAMOS NATIONAL LABORATORY, LOS ALAMOS, NEW MEXICO

Los Alamos is a federally funded Research and Development Center with a fundamental goal to manage national security challenges through scientific excellence.

Eric Brown.   Eric is currently the Division Leader for the Explosive Science and Shock Physics Division at Los Alamos National Laboratory where he oversees the premier research program on energetic materials and dynamic material response in support of national security. His research has spanned fracture and damage of complex heterogeneous polymers and polymer composites for energetic, reactive, and structural applications including crystalline phase transitions, plasticity, dynamic loading conditions, and self-healing materials. Within SEM he is the founding Editor-in-Chief of the *Journal of Dynamic Behavior of Materials*, served three terms as a Technical Editor for *Experimental Mechanics*, served as a Board of Directors as Member-at-Large, was a member of the Technical Activities Council, chaired the Research Committee and the Biological Systems and Materials Technical Division, is active in the Dynamic Behavior of Materials Technical Division, is a member of SEMEF, and received the SEM JSA Young Investigator Award. He received a B.S. in mechanical engineering and a M.S. and Ph.D. in theoretical and applied mechanics from the University of Illinois at Urbana Champaign. Eric received the SEM Durelli Award and became an SEM Fellow in 2017.

CHAPTER 8

SEM's Current Organizational Structure

In the preceding sections we have traced the beginnings of SESA/SEM and recognized a few of the many individuals that have created and developed the Society into what it is today. We have seen the connections between the successive generations that have carried the torch ahead toward the goal envisioned by its creators. The current structure of the Society includes the following Technical Divisions:

- Biological Systems and Materials

- Composite, Hybrid & Multifunctional Materials

- Dynamic Behavior of Materials

- Dynamic Substructures (An IMAC TD)

- Dynamics of Civil Structures (An IMAC TD)

- Fracture and Fatigue

- Inverse Problem Methodologies

- MEMS and Nanotechnology

- Modal Analysis/Dynamic Systems (An IMAC TD)

- Model Validation & Uncertainty Quantification (An IMAC TD)

- Nonlinear Structures and Systems (An IMAC TD)

- Optical Methods

- Sensors and Instrumentation (An IMAC TD)

- Thermomechanics and Infrared Imaging

- Time-Dependent Materials

It is interesting to see the comprehensive breath of topics covered by the different divisions and the profound evolution of the early structure that covered specific tools of research.

SEM Conferences.   SEM/SESA originally organized two annual conferences, the Spring conference and a topical conference in the Fall. The Fall conference is now dedicated to the field of Digital Image Correlation (DIC). Every four years the Spring conference becomes an International Congress where many more foreign members participate. Since 1991, SESA/SEM took over management of a new winter conference called IMAC.

8.1   THE IMAC STORY...*IT'S NOT JUST MODAL ANYMORE...*

The International Modal Analysis Conference (IMAC) was created by Dick DeMichele and Peter Juhl. The first conference was held in 1982. Until 1986, the conference was sponsored by Union College. From 1987–1991, the conference was co-sponsored by Union College and SEM. In 1991, SEM assumed total responsibility for the organization and management of IMAC and Union College agreed to the role of consulting partner. In 1996, Union College ceased any participation with IMAC.

During the initial planning for IMAC, the organizers quite understandably sought the support and advice of the leading United States and international individuals working in the field of modal testing and analysis.

These selected individuals then became the IMAC Advisory Board in 1982. Their function was to provide the IMAC staff with advice, suggestions, and support in the planning, promotion, and administration of the technical program for the annual IMAC Conference. The prestige and recognition of the IMAC Advisory Board by the modal community has been a major factor in the continued success of IMAC. Since 1982, the recognition and support given to IMAC by the universities, industry, and government agencies as the foremost international Conference devoted to modal testing and analysis has also been tantamount to its success.

In 1995, Dick DeMichele resigned as IMAC Technical Director and Professor Al Wicks, of Virginia Polytechnic Institute and State University, who had worked closely with Dick for a number of years assumed the position. Currently, Professor Al Wicks, Technical Director from Virginia Tech, works closely with Dr. Raj Singhal from the David Taylor Laboratory of the Canadian Space Agency, and the IMAC Advisory Board in planning the IMAC Conference.

SEM provides the financial backing and administrative staff for organizing and promoting the IMAC conferences. Staff is directly responsible to the SEM Executive Board for the prudent financial management of IMAC.

Over the past few years, IMAC has evolved by assuming areas outside the subject of Modal Analysis. The technical divisions that are part of the IMAC community are: Dynamics of Civil Structures, Dynamic Substructures (new in 2018), Modal Analysis/Dynamic Systems, Model Validation and Uncertainty Quantification for Structural Dynamics, Nonlinear Structures and Systems (new in 2018), and Sensors & Instrumentation. Our IMAC TD structure now contains five TDs other than Modal Analysis and continues to grow with the addition of Focus Groups and member activity.

In 2018, Professor Wicks wrote the following as his introduction in the IMAC Final Program...

IMAC has become a comprehensive meeting on a broad spectrum of technologies related to structural dynamics. Technologists and researchers will find presentations, tutorials, and products of interest. We link a technical agenda with an exposition that features a wide variety of products and services related to structural dynamics. In addition to the tutorials, short courses are available prior to the conference, making this conference a comprehensive event and value-added opportunity to build professional careers.

One of the unique attributes of IMAC, nurtured over the years, is the mix of analytical and experimental topics, bringing the analyst and the experimentalist together as a team. The traditional barriers have been removed to foster constructive dialogue between academics, industry and the government labs. It is from these meetings, that technologies are shared, enhancing our industries, infrastructure, our educational endeavors and improving society in general. As lofty as it sounds, IMAC remains a friendly conference where exhibitors, presenters and attendees spend several days exchanging the ideas that fuel the coming year.

We welcome all to IMAC to share in the vision of Dick DeMichele.

SEM's Annual conference focuses on the other technical divisions listed above: *Biological Systems and Materials*; *Composite, Hybrid & Multifunctional Materials*; *Dynamic Behavior of Materials*; *Fracture and Fatigue*; *Inverse Problem Methodologies*; *MEMS and Nanotechnology*; *Optical Methods*; *Residual Stress*; *Thermomechanics and Infrared Imaging*; *Time-Dependent Materials*. Recently, SEM Annual has added a conference track on Additive and Advanced Manufacturing with a focus on experimental mechanics. Both conferences look to keep their finger on the pulse of emerging areas of interest to the membership. The SEM Annual Conference is also the time of the year when all of the councils and committees meet to conduct the business for the year.

Journals. SEM publishes three peer-reviewed journals, *Experimental Mechanics*, the flagship of SEM, and *Experimental Techniques*, a publication to present more technically and applied topics of experimental mechanics of interest to industry and the membership. More recently, SEM launched the *Journal of Dynamic Behavior of Materials*, devoted to the science and engineering of material and structural response to dynamic loading, emphasizing high strain-rate, impact, blast, penetration, and shock response.

Proceedings. SEM publishes proceedings of the papers presented at the Society's Conferences.

Manuals and Monographs. SEM publishes manuals and monographs of interest to the community of experimental mechanicians.

In 2017, SEM launched a new publication with Morgan and Claypool publishers: *SEM Synthesis Lecture Series*. It is interesting to point out that no other professional engineering/scientific society that is as small as we are publishes three peer-reviewed journals plus conference proceedings, book series, etc.

SEM celebrates and encourages publishing its member's work...a means of sharing the intellectual equity/property of its members to continue the growth of the field of experimental mechanics.

SEM Membership.   Membership consists of approximately 1,500 members. In 2017, data showed 942 individual members, 322 student members, 61 corporate members, 10 honorary members, 82 lifetime members, and 142 retired Members.

Governance.   Following the initial tradition of SEM/SESA, the Society has a President, a Vice President, and an Executive Board. The Executive Board is formed by elected members, some appointed members, and two very important components. A permanent Executive Director/Secretary, a paid position, and a Treasurer. The Executive Secretary is a very important position because it provides continuity to the Society's governance. This was the position that William Murray held in honorary form for the Society's formative years.

The position of Executive Director/Secretary torch was passed from William Murray and Executive Committee and then to SESA/SEM's first Managing Director, Bonney Rossi who served the post from (1960–1979). The torch was then passed to Ken Galione from (1979–1996). From 1996–2000, there were a couple of staff members that served the post. From 2000–2002, former SEM President, Dr. Elizabeth (Beth) Stelts served as interim Executive Director with significan support from Dr. Jon Rogers who had just served his term as SEM President (2000–2001). Both Beth and Jon managed the search process that hired Dr. Thomas (Tom) W. Proulx as SEM's new Executive Director in 2002.

Tom brought new energy and an insatiable curiosity for learning of the past, present, and potential future for SEM. His background as a chemist and employee of PerkinElmer, served him well in both managing the day-to-day operations of SEM, and in finding areas for strategic growth. Tom worked closely with Jon Rogers to write a new *Experimental Mechanics* publishing contract with Springer Publishing giving SEM, for the first time, a more secure revenue stream from our flagship journal.

Tom served the society from 2002–2014, and his legacy lives on as SEM strives to achieve his vision, and that of William Murray, of a sustainable society.

Currently, the position of Executive Director and Secretary is held by Dr. Kristin Zimmerman. She was the inaugural SEM Student Paper Competition winner under the guidance of Professor Gary Cloud at Michigan State University in 1990; Chair of the Education Committee from 1991–2007; Associate Editor of *Experimental Techniques* from 1996-today, and Senior Editor from 2000–2007; President from 2008–2009; Assistant Treasurer 2012–2013; and appointed Treasurer in 2014. She was awarded the Tatnall award in 2014.

Kristin's professional career began with the General Motors (GM) Research and Development (R&D) Center in 1993–1997 where she created GM's Academic Partnerships program of over 100 Research Laboratories across the globe. From 1997–1999, Zimmerman worked in the areas of advanced engineering and design and in 1999/2000, she received a Fellowship to the National Academy of Engineering to work on STEM policy. From 2000–2009, Zimmerman worked in energy and environmental policy including an assignment in Beijing, China (2008–2009) managing GM China's Automotive Energy Research Center (CAERC) at Tsinghua University. She continued her energy and environmental policy work on the Chevy Volt Team, from 2006–2012. She is currently also President of MedFor: Inc., a translational sciences consulting firm spanning forensic medicine and engineering mechanics, founded with her husband in 1999.

Kristin's educational background includes: Physics, Mechanical Engineering, and Engineering Mechanics. She holds a Ph.D. in Engineering Mechanics from Michigan State University.

The basic structure of the Society has remained consistent throughout its existence. Sustainability of the Society and its structure was always of keen concern to the Society's father, William Murray. Unlike other professional societies, like the Societies of Civil Engineers, of Mechanical Engineers, or SPIE that involve large groups of members tied to large enterprises, and business interests and hence have very large memberships and financial resources, SEM has a reduced membership comparatively. William Murray was always concerned about the balance of academics and active industrial participants. This concern has continued to be part of SEM history, to increase the membership and to attract more actual applications-oriented members working in the industry. It is interesting to observe that in the history of SESA/SEM the number of conference attendees has been what it is today, about 550. Increasing the membership continues to be a goal for SESA/SEM. However, the steady membership has not been an obstacle for the relevant role that SESA/SEM has played in the U.S. and worldwide. This is proven by the growth of the SESA/SEM Conference attendance.

SESA/SEM has been the Society where many of the great advances in what, in general, has been called Experimental Mechanics has been introduced and the successive steps to reach applied technological levels have been part of discussion forums in the SESA/SEM meetings and publications. Starting with the roots of SEM/SESA when photoelasticity was the only field technique available, the birth of the most versatile and practically universal tool, the electrical strain gage SESA/SEM meetings and publications, were the vehicles were basic developments and actual procedures and necessary instrumentation and actual techniques of utilization were introduced and became available to the community. The same can be said with respect to the stress-coat, the next field technique available to get the location of strain gages. Photoelastic coatings came after the stress coat method, the tool for the field of stress-strain determination. This trend was followed with the Moiré methods, many of the applications of holographic interferometry to engineering problems as well as the different methods of speckle techniques, and digital image techniques to process fringe patterns. Image correlation techniques are one of the more recently applied methods of field analysis of displacements and its derivatives.

In the history of SESA/SEM we have seen how all this formidable array of experimental techniques were born in academic environments or research center groups, transmitted from creators to their collaborators and in many cases to their students creating a continuous chain of progress. The secret behind this remarkable success, despite the limited financial basis, is the dedication of the involved individuals. Those individuals that engaged in the preparation and organization of meetings and presentation preparation, know very well how many hours of their personal life and that of their collaborators and students go to these multiple tasks...tasks that many times dictate the annual rhythm of a group to be ready for a previously promised presentation or paper. A professional society, like SEM, is small but the growth of very valuable services from the Society deserves all the support of its membership and the community in general.

CHAPTER 9

Summary

What is the future? Today, in the technological world there is a powerful tendency toward the digitalization of all technological tasks. Companies reduce their production cycle by minimizing the time of the actual experimental verification of the compliance of their products to design goals and push for the digital verification compliance of their products. Remembering the times when many colleagues thought of the demise of the SESA/SEM because of the development of Finite Element Methods, and thinking of the enormous progress of artificial intelligence, one can conclude a similar fate for SEM. Again, with insight gained to the whole historical process of growth and development of SEM, we can be optimistic that by adapting and changing to provide support to new frontiers of progress, the survival of SEM is insured and constantly evolving with a new and more relevant area of technology focus pertinent to the corresponding time.

SEM's small size makes it nimble enough to indeed keep its finger on the pulse of what is new, exciting, and next. It can quickly embrace and grow those emerging areas that other professional societies are unable to due to their large bureaucratic structures. One such area for SEM is the area of Additive and Advanced Manufacturing (AAM). Yes, ASME and other professional organizations include AAM, but they are not as focused as SEM regarding the mechanics component of AAM. SEM can maintain this focus and continue to grow this area.

This is a living document that invites contributions from all SEM/IMAC members…

APPENDIX A

Chuck Taylor (1924–2017): Celebration of Life Stories by Michael Sutton

A.1 MICHAEL A. SUTTON STORIES

Chuck Taylor Memorial Celebration of Life
2/14/2018, Oak Room, Oak Hammock, Gainesville, FL
4pm to 5:30pm

First, I want to thank Oak Hammock for providing this opportunity to gather with family and friends today. I also want to thank our emcee Don Martin, Chuck's sons, Glenn and Gary Taylor, and their families for allowing me to be part of this celebration.

As many of you know, I was part of the Taylor's Series and graduated with my Ph.D. in Theoretical and Applied Mechanics at UIUC in the summer of 1981. In fact, there were three of us that graduated on the same day in June as part of the last cohort of Chuck's Ph.D. students at UIUC: myself, Frederick Mendenhall, and Albert Kai-Wai Wong. My peers at UIUC included Fred and Albert, as well as Bill Chao, and the incomparable Michael A. Tafralian, also known by all of us as the Mad Armenian. Mike Tafralian was a good friend of mine, and he taught me a great deal about life, since his interests were more about living life to its fullest and not about his research work. It was one of Chuck's characteristics that he was able to work with all sorts of students, including folks like Mike who took over 10 years to complete his degree!

Throughout my time at UIUC, I was guided by Chuck and his many friends on the faculty, including George Costello, Bob Miller, Jim Phillips, Don Carlson, Henry Langhaar, and Marvin Stippes in TAM and Prof. Langenbartel in Mathematics. Chuck was always looking out for his students, and his many friends took every opportunity to provide guidance and encouragement to me whenever needed.

A.2   CHUCK AND GOLF: A MATCH NOT MADE IN HEAVEN

One of Chuck's most memorable stories, and one he recounted every time we met, was about our golfing outing in 1979 at The Savoy, UIUC's golf course at that time. Now, I was a pretty good golfer in high school and received a full scholarship for golf at Southern Illinois University in 1968. I am not sure how Chuck found this out, but one day at UIUC he asked me if I would be interested in an advisor-student golf tournament. As his sons indicated earlier, Chuck was competitive and so I assumed he saw the opportunity to join forces with me and beat his competition. In this case, his competition was George Costello, who had already agreed to be a part of the tournament with one of his Ph.D. students. Since it was a handicap tournament, I asked Chuck what kind of golfer he was, and he said that he "played DOUBLE BOGEY golf." To all golfers, this means that you play each hole two over par. Thus, for 18 holes, this means you would be given 36 shots as your handicap. Well, I got excited at this, because I was playing with a 6 handicap (shooting 78 on a par 72 course) and felt that Chuck could really help us on several holes since he had all those strokes of handicap. So, I agreed to join him and, on the day the tournament started, Chuck, myself, George Costello, and his Ph.D. student were in a foursome. I teed off first and hit my drive down the middle of the fairway. Chuck walked up on the tee, put his ball on a tee and took out an iron instead of a driver; I guess that I should have known something was amiss when I saw this… Chuck then took his first swing and missed the ball entirely, taking a large chunk of grass behind the ball. Mumbling something, he took another swing and topped the ball, rolling it gently about 30 feet down the tee. At this point, I knew something was off. Sure enough, the entire first nine holes was a repeat of the initial swing…, hacking, topping, missing entirely continued unabated. Occasionally, Chuck just pick up his ball and did not finish the hole since he had swung so many times. Eventually, we finished the first nine holes, with me shooting a 39 and Chuck shooting something above 70, which was at least 34 shots over par in just nine holes! When we finally finished the entire 18 holes, Chuck had not helped on any hole and we lost by several shots to the happy, smiling George Costello team. Frustrated, I said to Chuck after the round, "Chuck, I thought you said you were a DOUBLE BOGEY golfer!" With an impish smile, Chuck said with a straight face, "Yes, I am. If BOGEY is a 5 on a par 4 hole, then I shoot "DOUBLE" BOGEY, or 10!!!" At that point, I guess my face showed a stunned look, as I realized Chuck knew all along that we never did have a chance. Closing out the day on the appropriate note, George Costello said brightly, "I guess the drinks are on you guys!"

A.3   A FISHY STORY

As we heard earlier, Chuck really loved fishing and he was quite good at it. He regularly caught big fish and traveled the world, fishing with friends and colleagues. One of his fishing buddies was Prof. Arthur P. Boresi. Prof. Boresi left UIUC in the late 1970s becoming chairperson at the University of Wyoming, which is truly God's country for outdoorsmen and fishermen.

Chuck visited Art several times, fishing with him. It was after one of these trips that Chuck told me about catching an 8-pounder, spreading his arms wide to show the size of the fish. Just as Chuck was finishing his story, a colleague entered the room and heard the end of the story. Well, I heard a chuckle and turned to see Chuck's colleague shaking his head, saying "Chuck, every time I come in here, the length of the fish gets longer and the weight gets bigger. If you are not careful, the next time you tell the story, you may need to get longer arms!" As I turned to look at Chuck, he had that same impish smile, saying "It's the story that matters, not the size of the fish!" After hearing this story, I always remembered it with a smile. However, Chuck never brought it up again!

A.4 A FRIEND FOR ALL SEASONS

Throughout my career, my friend Chuck Taylor was always around, quietly assisting me when he had the opportunity to do so. He never talked about what he did for me, but I have learned from others just how much he has done over the past 40 years. He helped me get my position at South Carolina in 1982. He helped me obtain the NSF Presidential Young Investigator Award from President Reagan in 1986. He introduced me to many of his friends in academia and SEM: Jim Dally, C.W. Smith, Bill and Michael Fourney, Albert Kobayashi and Daniel Drucker are a few of those that I have had the good fortune to know over the years. He helped me obtain the SEM CE Taylor Award in 2008, and I will always cherish the fact that he personally gave me the award. He provided guidance to me on both personal and professional issues whenever I called. Though Chuck is gone now, I will always remember his kindness, generosity of spirit, and willingness to assist his students, especially me. I have done my best to try and help others, in the same way Chuck helped me, and I will do my best to honor his memory by "Paying It Forward" whenever I have the opportunity to do so.

Figure A.1: Photo taken at Chuck Taylor's Celebration of Life Memorial: L-R, Ghatu Subhash, Bill Chao, Mike Sutton, Bill Fourney, Peter Ifju, and Jim & Anne Dally.

APPENDIX B

Past and Present SEM Officers (1934–2018)

Please see the tables on the pages that follow.

	President	Executive Committee	Secretary–Treasurer	Executive Committee	
1943–1944	W. M. Murray	C. Lipson	M. Heyényi	R. D. Mindlin	
	President	**Vice President**	**Secretary–Treasurer**	**Executive Committee**	**Executive Committee**
1944–1945	M. Heyényi	C. Lipson	W. M. Murray	R. D. Mindlin	C. O. Dohrenwend
1945–1946	C. Lipson	R. D. Mindlin	W. M. Murray	C. O. Dohrenwend	O. J. Horger
1946–1947	R. D. Mindlin	C. O. Dohrenwend	W. M. Murray	E. L. Shaw	E. K. Timby
	President	**Vice President**	**Vice President**	**Secretary–Treasurer**	**Executive Committee**
1947–1948	C. O. Dohrenwend	R. E. Peterson	E. L. Shaw	W. M. Murray	I. G. Hedrick
1948–1949	R. E. Peterson	E. L. Shaw	R. Weller	W. M. Murray	J. H. Meier
1949–1950	E. L. Shaw	R. Weller	J. H. Meier	W. M. Murray	T. J. Dolan
1950–1951	J. H. Meier	T. J. Dolan	J. Marin	W. M. Murray	M. Holt
1951–1952	T. J. Dolan	J. Marin	Walter Ramberg	W. M. Murray	D. C. Drucker
1952–1953	Walter Ramberg	J. Marin	Marshall Holt	W. M. Murray	W. V. Covert
1953–1954	Marshall Holt	S. S. Manson	M. M. Leven	W. M. Murray	C. W. Gadd
1954–1955	Joseph Marin	S. S. Manson	M. M. Leven	W. M. Murray	G. Herrmann
1955–1956	S. S. Manson	M. M. Leven	E. Wenk, Jr.	W. M. Murray	W. R. Cambell
1956–1957	M. M. Leven	E. Wenk, Jr	W. R. Cambell	W. M. Murray	H LaTour
1957–1958	E. Wenk, Jr	W. R. Cambell	B. J. Lazan	W. M. Murray	D. C. Drucker
1958–1959	W. R. Cambell	B. J. Lazan	D. C. Drucker	W. M. Murray	G. Ellis
1959–1960	B. J. Lazan	D. C. Drucker	I. Vigness	W. M. Murray	W. W. Soroka
	President	**Vice President**	**Vice President**	**Treasurer**	**Executive Secretary**
1960–1961	D. C. Drucker	I. Vigness	W. W. Soroka	W. M. Murray	B. E. Rossi
1961–1962	I. Vigness	W. W. Soroka	C. C. Perry	F. C. Bailey	B. E. Rossi

	President	Vice President	Vice President	Treasurer	Executive Secretary
1962–1963	W. W. Soroka	C. C. Perry	J. C. New	F. C. Bailey	B. E. Rossi
1963–1964	C. C. Perry	J. C. New	R. Guernsey	F. C. Bailey	B. E. Rossi
1964–1965	John C. New	R. Guernsey	C. E. Taylor	F. C. Bailey	B. E. Rossi
1965–1966	Roscoe Guernsey	C. E. Taylor	F. C. Bailey	F. C. Bailey	B. E. Rossi
1966–1967	C. E. Taylor	F. C. Bailey	C. S. Barton	A. E. Johnson, Jr.	B. E. Rossi
1967–1969	F. C. Bailey	C. S. Barton	J. W. Dally	A. E. Johnson, Jr.	B. E. Rossi
1969–1970	C. S. Barton	J. W. Dally	A. E. Johnson, Jr.	A. E. Johnson, Jr.	B. E. Rossi
1970–1971	J. W. Dally	A. E. Johnson, Jr.	C. E. Work	R. H. Homewood	B. E. Rossi
1971–1972	A. E. Johnson, Jr.	C. E. Work	R. C. A. Thurston	R. H. Homewood	B. E. Rossi
1972–1973	C. E. Work	R. C. A. Thurston	E. E. Day	R. H. Homewood	B. E. Rossi
1973–1974	R. C. A. Thurston	E. E. Day	J. W. Dalley	R. H. Homewood	B. E. Rossi
1974–1975	E. E. Day	J. W. Dalley	E. I. Riegner	R. H. Homewood	B. E. Rossi
1975–1976	J. W. Dalley	E. I. Riegner	D. R. Harting	R. H. Homewood	B. E. Rossi
1976–1977	E. I. Riegner	D. R. Harting	H. F. Brinson	R. H. Homewood	B. E. Rossi
1977–1978	D. R. Harting	H. F. Brinson	R. L. Johnson	R. H. Homewood	B. E. Rossi
1978–1979	H. F. Brinson	R. L. Johnson	M. E. Fourney	R. H. Homewood	B. E. Rossi
1979–1980	R. L. Johnson	M. E. Fourney	D. L. Willis	R. H. Homewood	K. A. Galione
1980–1981	M. E. Fourney	D. L. Willis	S. E. Swartz	R. H. Homewood	K. A. Galione
1981–1982	D. L. Willis	S. E. Swartz	J. B. Ligon	C. A. Calder	K. A. Galione
1982–1983	S. E. Swartz	J. B. Ligon	W. N. Sharpe, Jr.	C. A. Calder	K. A. Galione
1983–1984	J. B. Ligon	W. N. Sharpe, Jr.	R. J. Rinn	C. A. Calder	K. A. Galione
1984–1985	W. N. Sharpe, Jr.	R. J. Rinn	I. M. Allison	D. L. Willis	K. A. Galione
1985–1986	R. J. Rinn	I. M. Allison	C. A. Calder	D. L. Willis	K. A. Galione
1986–1987	I. M. Allison	C. A. Calder	S. K. Foss	S. E. Swartz	K. A. Galione
1987–1988	Clarence A. Calder	Susan K. Foss	Albert S. Kobayashi	Stuart E. Swartz	Kenneth A. Galione
1988–1989	Susan K. Foss	Albert S. Kobayashi	Robert F. Sullvan	Clarence Chambers	Kenneth A. Galione

	President	Vice President	Vice President	Treasurer	Executive Secretary
1989–1990	Albert S. Kobayashi	Robert F. Sullvan	W. L. Fourney	Clarence Chambers	Kenneth A. Galione
1990–1991	Robert F. Sullvan	W. L. Fourney	Frank D. Adams	Clarence Chambers	Kenneth A. Galione
1991–1992	W. L. Fourney	Frank D. Adams	Gary Cloud	Susan K. Foss	Kenneth A. Galione
1992–1993	Frank D. Adams	Gary L. Cloud	T. Dixon Dudderar	Susan K. Foss	Kenneth A. Galione
1993–1994	Gary L. Cloud	T. Dixon Dudderar	Mark E. Tuttle	Susan K. Foss	Kenneth A. Galione
1994–1995	T. Dixon Dudderar	Mark E. Tuttle	Elizabeth A. Fuchs	Robert F. Sullivan	Kenneth A. Galione
1995–1996	Mark E. Tuttle	Elizabeth A. Fuchs	Ravinder Chona	Robert F. Sullivan	Kenneth A. Galione
1996–1997	Elizabeth A. Fuchs	Ravinder Chona	Charles E. Harris	Robert F. Sullivan	Kristin L. MacDonald
1997–1998	Ravinder Chona	Charles E. Harris	Igor Emri	Robert F. Sullivan	Kristin L. MacDonald
1998–1999	Charles E. Harris	Igor Emri	Jonathan D. Rogers	Robert F. Sullivan	Kristin L. MacDonald
1999–2000	Igor Emri	Jonathan D. Rogers	Michael A. Sutton	Elizabeth A. Stelts	MacDonald/Deuschle(A)
2000–2001	Jonathan D. Rogers	Michael A. Sutton	Arun Shukla	Stelts/Previs (I)	Deuschle (A)/Stelts (I)
2001–2002	Michael A. Sutton	Arun Shukla	Randy Allemang	Sharon Previs (I)	Elizabeth A. Stelts (I)
2002–2003	Arun Shukla	Randy Allemang	Jose L. F. Freire	Jonathan D. Rogers	Tom Proulx
2003–2004	Randy Allemang	Jose L. F. Freire	Eddie O'Brien	Jonathan D. Rogers	Tom Proulx
2004–2005	Jose L. F. Freire	Eddie O'Brien	Masahisa Takashi	Jonathan D. Rogers	Tom Proulx
2005–2006	Eddie O'Brien	Masahisa Takashi	Archie Andonian	Jonathan D. Rogers	Tom Proulx
2006–2007	Masahisa Takashi	Archie Andonian	Kristin Zimmerman	Jonathan D. Rogers	Tom Proulx
2007–2008	Archie Andonian	Kristin Zimmerman	Wei-Chung Wang	Jonathan D. Rogers	Tom Proulx
2008–2009	Kristin Zimmerman	Wei-Chung Wang	R. Pryputniewicz	Jonathan D. Rogers	Tom Proulx
2009–2010	Wei-Chung Wang	R. Pryputniewicz	Peter Ifju	Jonathan D. Rogers	Tom Proulx
2010–2011	R. Pryputniewicz	Peter Ifju	Carlos Ventura	Jonathan D. Rogers	Tom Proulx
2011–2012	Peter Ifju	Carlos Ventura	Emmanuel Gdoutos	Jonathan D. Rogers	Tom Proulx
2012–2013	Carlos Ventura	Emmanuel Gdoutos	Nancy Sottos	Jonathan D. Rogers	Tom Proulx
2013–2014	Emmanuel Gdoutos	Nancy Sottos	Guruswami Ravichandran	Jonathan D. Rogers	Tom Proulx

	President	Vice President	Vice President	Treasurer	Executive Secretary
2014–2015	Nancy Sottos	Guruswami Ravichandran	Peter Avitabile	Jonathan D. Rogers	Kristin B. Zimmerman
2015–2016	Guruswami Ravichandran	Peter Avitabile	Kathryn Dannemann	Jonathan D. Rogers	Kristin B. Zimmerman
2016–2017	Peter Avitabile	Kathryn Dannemann	Wendy Crone	Jonathan D. Rogers	Kristin B. Zimmerman
2017–2018	Kathryn Dannemann	Wendy Crone	John Lambros	Jonathan D. Rogers	Kristin B. Zimmerman
2018–2019	Wendy Crone	John Lambros	Daniel Rixen	Jonathan D. Rogers	Kristin B. Zimmerman

Authors' Biographies

CESAR A. SCIAMMARELLA

Cesar A. Sciammarella is Professor Emeritus in the Mechanical, Materials, and Aerospace Department at Illinois Institute of Technology. For 37 years he was the Director of the world-renowned Experimental Mechanics Laboratory of the Illinois Institute of Technology. He is currently an Adjunct Professor at the Department of Mechanical Engineering of the Northern Illinois University. From 1991–1998, Cesar was a Non-Resident Professor at the Universita Degli Study, in Nuoro, Italy. He was Visiting Professor at the following institutions: Polytechnic Institute of Milano (1972, 1976); University of Cagliari (1979); Polytechnic Institute of Lausanne (1979); Univeristy of Poitiers (1980); and Polytechnic Institute of Bari (1992, 1994, 1998, 2003–2008). At the Polytechnic Institute of Bari, he completed a five-year project funded by a personal grant from the Italian government to broaden and revamp the Experimental Mechanics laboratory.

After graduation from college, Cesar continued his Ph.D. on the Moiré method resulting in the first computer-supported processing of fringe patterns in 1968. He has continued this work in fringe pattern information decoding arriving to self-consistent computer-based algorithms that overcame most of the limiting factors that put constraints on other approaches guaranteeing maximum accuracy and computer efficiency. He applied these different developments to Moiré patterns, speckle patterns, holographic patterns in 2D and 3D applications, and MRI patterns. These applications cover a wide spectrum of scales, macro, micro, nanometer, sub-nanometer scales ,and the levels of atomic structures. In 2006, he introduced the detection of optical signals beyond the Rayleigh limit extending holography information recovery at the nano range. Besides working on developing optical techniques applied to experimental mechanics, he has utilized these techniques and other methods of experimental mechanics to solve important engineering problems in the areas of: fracture mechanics and fatigue of metals; fracture of fiber reinforced composites; and fracture of particle reinforced metals. He has done work on bio-mechanics in such areas as damage in knee-cap replacement and problems of the stability of the vertebral column, as well as the analysis of MRI records to obtain heart displacements and strains. He has also worked in wheel and rail contact problems, and more recently has done work in laser-adding manufacturing. In addition, he participated in two important projects of the U.S. space program: the Apollo IV and the Space Shuttle program.

Cesar has authored more than 350 papers and book chapters. His textbook *Modern Experimental Mechanics of Solids: Theory, Techniques, Instrumentation and Applications* was published by Wiley in 2012.

He has received many SEM awards, including the, Hetenyi, Lazan, and Frocht awards and the William M. Murray Medal. In 2011, he received the Theocaris Award from the Society for Experimental Mechanics in recognition of his pioneering efforts in developing and applying Moiré, holography, speckle interferometry methodologies, and microscopic versions of these techniques to Experimental Mechanics. In 2013, he was awarded the Honorary Membership of the Society for Experimental Mechanics—the highest honor the society awards. In 2011, the British Society for Strain Measurement awarded him and his co-workers the Fylde Electronic Price for best paper published in *Strain* in 2010. In 2013, a symposium titled Advances in Experimental Mechanics for Industrial Design and Automation was organized at the Polytechnic of Bari to honor his 90th birthday. It was at this symposium that he was presented with a lifetime achievement award. In 2014, the International Conference on Computational and Experimental Engineering and Sciences, at a meeting in Changwon, Korea, also presented him with a lifetime achievement award and in 2016 he received yet another lifetime achievement award, this time from the European Society for Experimental Mechanics at a meeting in Rhodes, Greece.

He is a lifetime member of the American Society of Mechanical Engineers and an honorary lifetime member of the Italian Association of Stress Analysis.

KRISTIN B. ZIMMERMAN

Dr. Zimmerman's professional career began with the General Motors (GM) Research and Development (R&D) Center in 1993 where she created GM's Satellite Laboratory Network establishing over 100 GM Satellite Research Laboratories across the globe. In 1999, Zimmerman received a Fellowship to the National Academy of Engineering to work on STEM policy. From 2000–2012, she worked in the areas of advanced engineering and design, energy and environmental policy, and was responsible for creating and implementing GM's energy and carbon management strategies across the globe. From 2008–2009, Zimmerman lived in Beijing, China managing GM China's Automotive Energy Research Center (CAERC) on the campus of Tsinghua University. She continued her energy and environmental policy work as an original member of the Chevy Volt Team, 2006–2012.

Dr. Zimmerman is currently the Executive Director of the Society for Experimental Mechanics and a 30-year member of the Society. Her educational background includes: Physics, Mechanical Engineering, and Engineering Mechanics. Her Ph.D. is in Engineering Mechanics from Michigan State University.

Additional Notes